Homework for MONDAY 2/24/86

Foothill-DeAnza
O. T. I.
12345 El Monte Road
Los Altos Hills, CA 94022

① Chapter 5 — pag 100 & 101
All the problems,
except # 4, 18, 19, 20

② LAB # Report : First Mask
SUBSTRATE PREP TO DEV CHECK

Modern semiconductor fabrication technology

Peter Gise
  Tencor Instruments, Mountain View, CA

Richard Blanchard
  Siliconix, Santa Clara, CA

A Reston Book
Prentice-Hall
Englewood Cliffs, New Jersey 07632

Library of Congress Cataloging in Publication Data

Gise, Peter E.
  Modern semiconductor fabrication technology.

  Includes index.
  1. Semiconductors—Design and construction.
  2. Integrated circuits—Design and construction.
  I. Blanchard, Richard. II. Title.
  TK7871.85.G48 1986   621.3815'2   85-18252
  ISBN 0-8359-4684-3

© 1986 by Prentice-Hall
Englewood Cliffs, NJ 07632

A Reston Book
Published by Prentice-Hall
A Division of Simon & Schuster, Inc.
Englewood Cliffs, NJ 07632

All rights reserved. No part of this book may be reproduced in any way or by any means without permission in writing from the publisher.

10 9 8 7 6 5 4 3 2 1

PRINTED IN THE UNITED STATES OF AMERICA

Contents

Preface vii

1 Chemistry and physics of semiconductor materials 1

    1.0   Introduction, 1
    1.1   Atomic Structure, 1
    1.2   Classification of Materials, 4
    1.3   Doping of Semiconductors, 7
    1.4   Resistivity of a Semiconductor Doped with a Single Impurity, 9
    1.5   Resistivity of a Semiconductor Doped with Both n-type and p-type Impurities, 12
    1.6   Carrier Transport, 14
    Review Exercises, 16

2 Crystal growth and wafer preparation 19

    2.0   Introduction, 19
    2.1   Origin of Silicon and Its Purification, 20
    2.2   Crystal Growth, 22
    2.3   Doping of Crystals During Growth, 24
    2.4   Wafer Manufacture, 25
    2.5   Crystal Orientation, 27
    2.6   Crystal Defects, 29
    Review Exercises, 31

3 Oxidation of silicon 33

    3.0   Introduction, 33
    3.1   The Growth of Silicon Dioxide, 33
    3.2   Equipment for Thermal Oxidation, 35
    3.3   The Oxidation Process, 35

    3.4   Oxide Evaluation, 40
    3.5   Oxidation Technology for Low Contamination Levels, 41
    3.6   Oxidation Reaction, 42
    3.7   Oxide Thickness Determination, 43
    3.8   Redistribution of Dopant Atoms During Thermal Oxidation, 46
    3.9   High-Pressure Oxidation, 47
    3.10 Anodic Oxidation, 48
          Review Exercises, 49
          References, 50

4 Photolithography   51

    4.0   Introduction, 51
    4.1   Process Overview, 51
    4.2   Photoresist: The Key to Image Transfer, 53
    4.3   The Photolithographic Process Sequence, 54
    4.4   Photomasks, 66
          Review Exercises, 69

5 Impurity introduction and redistribution   71

    5.0   Introduction, 71
    5.1   Definition of Diffusion, 71
    5.2   The Diffusion Process, 72
    5.3   Diffusion Analysis, 83
    5.4   Mathematics of Diffusion, 83
    5.5   Ion Implantation, 94
    5.6   Mathematics of Ion Implantation, 94
    5.7   Limitations of Ion Implantation, 100
          Review Exercises, 100
          References, 101

6 Epitaxial deposition   103

    6.0   Introduction, 103
    6.1   The Epitaxial Reactor, 105
    6.2   The Epitaxial Growth Sequence, 112
    6.3   Evaluation of Epitaxial Layers, 117
          Review Exercises, 120
          References, 120

7 Nonepitaxial chemical vapor deposition   123

    7.0   Introduction, 123
    7.1   CVD Methods, 126
    7.2   CVD Procedures and Uses, 128
          Review Exercises, 133

Contents

8 Metallization 135

- 8.0 Introduction, 135
- 8.1 Metallization Requirements, 135
- 8.2 Vacuum Deposition, 136
- 8.3 Deposition Techniques, 142
- 8.4 Vacuum Deposition Cycle, 145
- 8.5 Evaluation of Film Characteristics, 147
- Review Exercises, 147
- References, 148

9 Device processing: from alloy to sale 149

- 9.0 Introduction, 149
- 9.1 Alloying—Annealing, 149
- 9.2 Post-Alloy Sample Probing, 152
- 9.3 Scratch Protection, 153
- 9.4 Back-Side Preparation, 154
- 9.5 Wafer Sort, 155
- 9.6 Device Separation, 155
- 9.7 Die-Attach (Die Bonding), 156
- 9.8 Wire Bonding, 156
- 9.9 Packaging Considerations, 157
- 9.10 Final Test, 157
- 9.11 Mark and Pack, 157
- Review Exercises, 158

10 Device and IC technologies 159

- 10.0 Introduction, 159
- 10.1 Bipolar Technology, 159
- 10.2 Devices Fabricated Using Standard Bipolar Technology, 161
- 10.3 Basic MOS Technology, 168
- 10.4 MOS Technology Variations, 170
- 10.5 Comparison of Bipolar and MOS IC Technologies, 172
- Review Exercises, 172

11 The wafer fabrication environment 175

- 11.0 Introduction, 175
- 11.1 Chemicals and Cleaning Procedures, 177
- 11.2 Water, 178
- 11.3 Air, 181
- 11.4 Gases, 182
- 11.5 Particle-Monitoring Technology, 182
- 11.6 Personnel and Clean Room Procedures, 183
- Review Exercises, 184

12 Semiconductor measurements 185

12.0 Introduction, 185
12.1 Evaluation of Semiconductor Materials, 185
12.2 Evaluation of Diffused Layers, 188
12.3 Diffusion Profile Measurements, 189
12.4 Epitaxial Layer Evaluation, 193
12.5 Oxide and Thin-Film Evaluation Techniques, 196
12.6 Ion Implant Evaluation, 203
12.7 Metallization Monitoring, 205
Review Exercises, 206
References, 206

13 Advanced silicon technology 207

13.0 Introduction, 207
13.1 Dominant Trends in Technology: Substrate Size and Device Density, 207
13.2 Alignment-and-Exposure Step, 208
13.3 Developments in Other Processing Areas, 211
13.4 Developments in Device Technology, 211
Review Exercises, 214

14 Nonsilicon technology 215

14.0 Introduction, 215
14.1 Light-Emitting Diodes and Laser Diodes, 215
14.2 Optical Integrated Circuits, 216
14.3 Liquid Crystal Displays, 216
14.4 Gallium Arsenide Transistors and ICs, 217
14.5 Josephson Junctions, 218
14.6 Quartz Crystal Oscillators, 218
14.7 Magnetic Bubble (Magnetic Domain) Devices, 218
14.8 Hybrid Technology, 219
Review Exercises, 220

Appendices 221

1 Scientific notation, 221
2 Use of graphs, 227
3 Units, 233
4 Solutions, 237
5 Glossary, 253

Index 263

Preface

The purposes of this text are to provide a single source of reference to those individuals involved in the processing of semiconductors and to introduce students of other technologies to the technology of modern semiconductor processing. The text arose from a series of college courses for the semiconductor technician and was expanded to include many aspects of process design of interest to the processing engineer.

Although this text will be of most interest to the process engineer and technician, it should also prove valuable to the circuit designer and device engineer. The process information will supply useful background information about the manner in which the properties of the finished devices are related to specified process parameters.

A wide range of material is presented and it is arranged in approximately the same order as the actual process flow during the manufacturing process. Additional material has been added covering semiconductor measurements.

1 | Chemistry and physics of semiconductor materials

1.0 | Introduction

This chapter provides information on the materials used in semiconductor process technology. A basic explanation of atomic structure is used as the starting point. Next, the periodic table of the elements is used as a foundation for understanding the chemical and electrical behavior of these materials. Finally, key electrical concepts such as conductivity, resistivity, and mobility are discussed.

1.1 | Atomic Structure

Early structural models of the atom pictured it as having a nucleus containing both positively charged protons and electrically neutral neutrons surrounded by orbitals or shells containing negatively charged electrons (Figure 1–1). This model of the atom is being continuously refined by atomic physicists, but the features of the model are sufficient to explain much of the physical behavior observed in materials, including most semiconductors.

An atom in its natural or electrically neutral state has the same number of electrons and protons. However, the gain or loss of electrons from the orbitals surrounding the nucleus produces an atom that is charged either positively or negatively. An atom charged in such a fashion is called an *ionized atom*, or an *ion*. The majority of the physical and chemical properties of an atom are determined by the number of electrons in its outermost orbital, since these electrons are the ones that interact with the outside world.

All atoms with the same number of protons (regardless of the number of

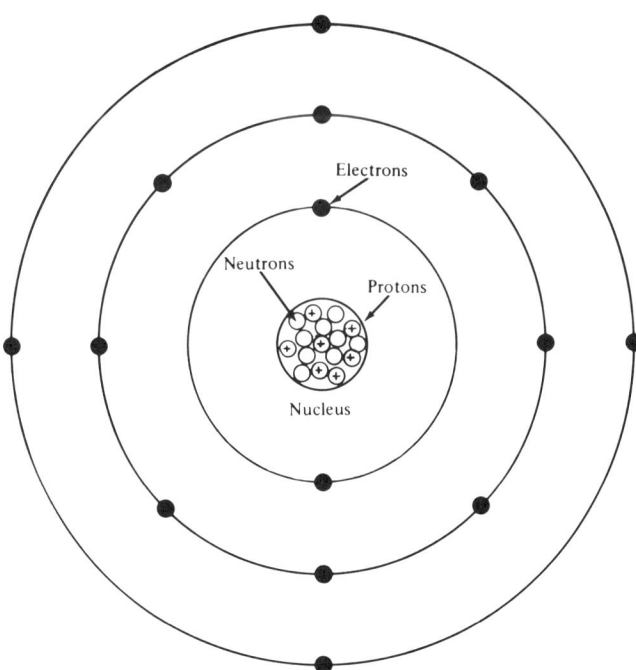

Figure 1–1: The atomic structure of a silicon atom.

neutrons or electrons) are the same element. Nonionized atoms with the same number of protons must also have the same number of electrons. Hence, only the number of neutrons contained in the nucleus can differ. Atoms with the same number of protons but a different number of neutrons are *isotopes* of the element.

Studies by several nineteenth century chemists detected similarities in the physical and chemical properties of elements having different densities. Groupings of elements with similar properties based on their densities led to the periodic table (Figure 1–2). This table was arrived at largely by experimental means, but it offers profound insights into the behavior of materials. The uncomplicated but accurate picture of semiconductors obtained through the use of the periodic table is sufficient for all but advanced semiconductor work.

The current periodic table (modified from Mendeleev's original) is based upon the arrangement of electrons around the nucleus of an atom. Every atom has orbitals or shells that electrons may occupy. The orbitals closer to the nucleus hold fewer electrons than the orbitals farther away. Electrons fill orbitals starting from the innermost. The rows of elements in the periodic table correspond to the filling of an orbital with electrons. When an orbital is filled, a new row in the periodic table is begun. Elements that are in the same column in the periodic table have the same number of electrons in the outermost or-

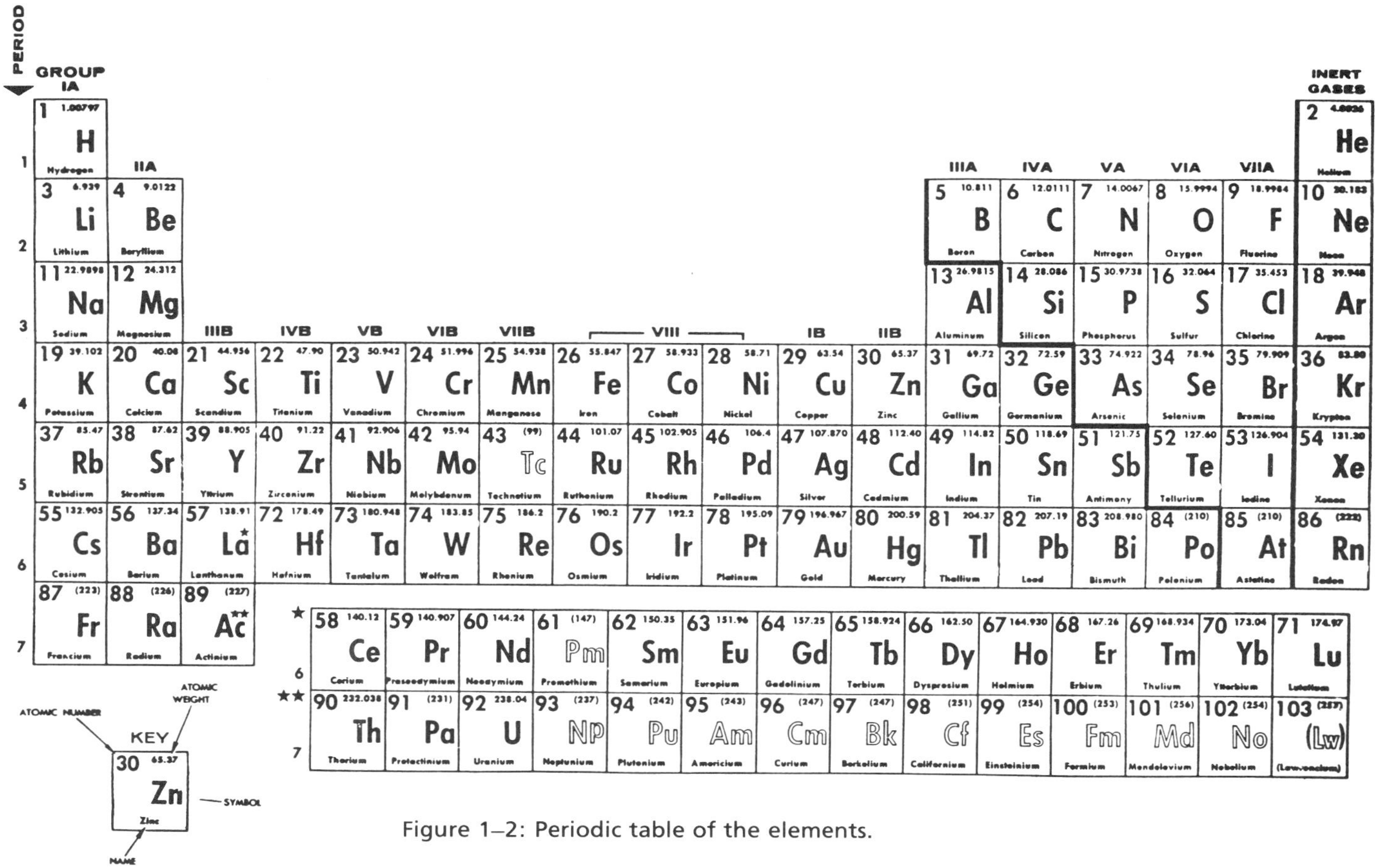

Figure 1-2: Periodic table of the elements.

bital. The columns are given "group numbers" that indicate the number of electrons in the outer orbital.

Dimitri Ivanovich Mendeleev, the Russian chemist who devised the periodic table, noted that atoms with eight electrons in their outer orbitals are chemically inert. The observation that atoms have a complete set of electrons when they have eight electrons in their outer shell explains the chemical reactions of elements to form compounds. Group I elements (with one electron in their outer shell) react with Group VII elements (with seven electrons in their outer shell). The Group VII atom "borrows" the electron from the Group I element to complete its outer shell. This borrowing leaves the Group I atom with no electrons in its outer shell, but with completed shells beneath. Each element now has a full complement of electrons in its outer shell. The atoms are held together by the electric force between the atom with one extra electron and the atom with one fewer electron. This type of bonding is known as *ionic bonding*.

In a similar manner, when a Group II and a Group VI element combine, each atom satisfies its need for electrons. However, the Group VI atom has difficulty capturing the two extra electrons, so it shares them instead. The bond formed is less ionic (electron-taking) and more covalent (electron-sharing). In like fashion, Group III atoms combine with Group V atoms, and Group IV atoms combine with Group IV atoms. A Group IV atom will share one of its four electrons with each of its four nearest neighbors part of the time and borrow one electron from these neighbors part of the time.

1.2 | Classification of Materials

One scheme scientists use to classify materials is to group them by their ability to conduct electricity. Three broad categorizations are:

1. Insulator—no current flows when a voltage is applied.
2. Metal—current flows easily when a voltage is applied.
3. Semiconductor—some current flows when a voltage is applied.

If we look at the electron structure of these three classes of materials (Figure 1-3), we see that:

1. Insulators have all their electrons tightly bound, so none are free to carry current.
2. Metals have many electrons readily available to carry current.
3. Semiconductors have some electrons free to carry current.

1.2 Classification of Materials

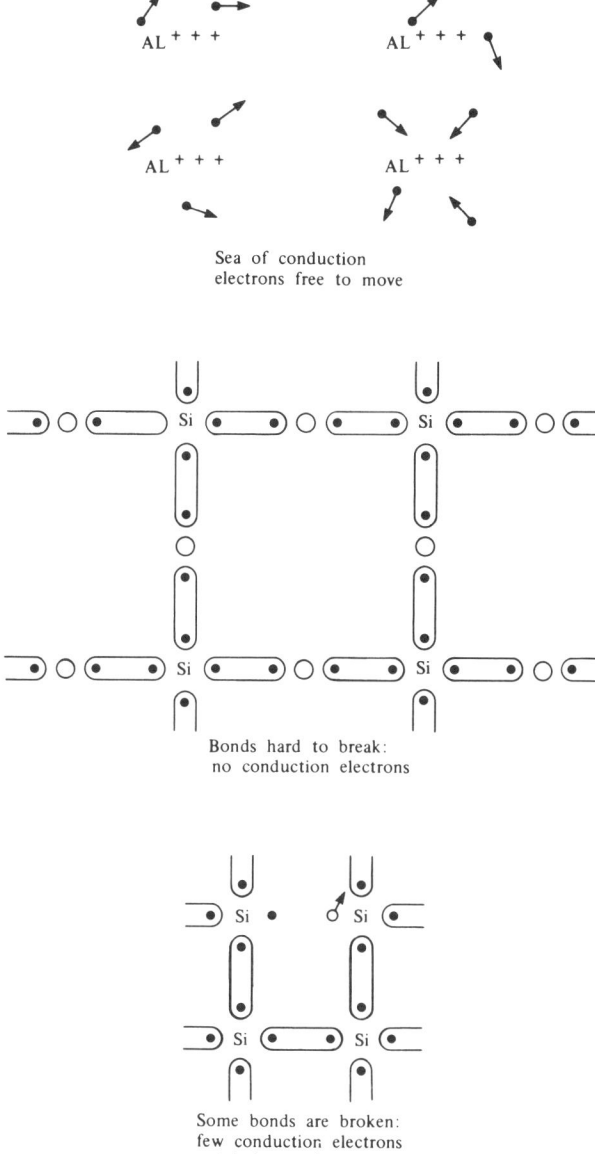

Figure 1–3: Bonding diagrams of a metal, an insulator, and a semiconductor.

Taking a closer look at semiconductors, of which silicon and germanium are two examples, we see that they both belong to Group IV in the periodic table. Thus, when these elements are crystalline, an atom shares one of its four electrons with its four nearest neighbors (Figure 1–4a). However, at any temperature greater than absolute zero (0 K), some of the bonds linking the atoms are broken (Figure 1–4b). The broken bonds produce electrons that are free to conduct electricity. In addition, any broken bonds corresponding to the absence of an electron are also free to move in the lattice (Figure 1–4c). (The absence of an electron is called a hole; the concept is similar to calling those volumes of, say, a stream in which water is absent a bubble.)

In a pure semiconductor crystal, the number of broken bonds depends only on the temperature. Since every broken bond produces both a hole and a conduction electron, both are present in equal numbers. The symbol n is used to signify the number of conduction electrons/cm^3 in a semiconductor, while

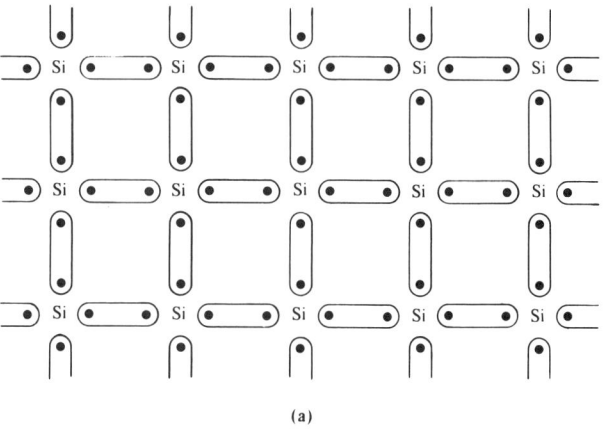

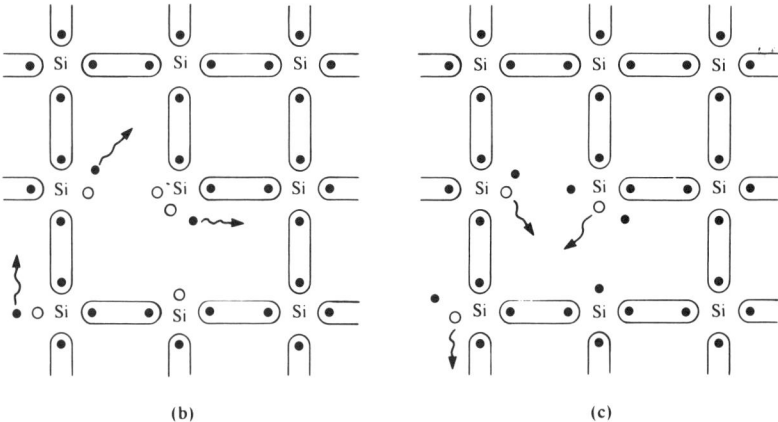

Figure 1–4: (a) Silicon at absolute zero; (b) Electron conduction in silicon; (c) Hole conduction in silicon.

the symbol p signifies the number of holes/cm³ in a semiconductor. Since n and p are equal in pure or "intrinsic" silicon, we can say that n equals p. The number of broken bonds in an intrinsic sample is called n_i, and it follows that

$$n = p = n_i \qquad (1\text{–}1)$$

and

$$n \cdot p = n_i^2 \qquad (1\text{–}2)$$

where n_i^2 depends only on temperature. In silicon at room temperature (27°C), $n_i = 1.4 \times 10^{10}/\text{cm}^3$ and $n_i^2 \cong 2 \times 10^{20}/\text{cm}^6$.

1.3 Doping of Semiconductors

The presence of equal numbers of holes and conduction electrons leads to no interesting phenomena, but the ability to increase the number of holes or conduction electrons by adding trace amounts of impurities, called *dopants*, means that regions of semiconductor materials may be altered to perform useful functions. Silicon has four electrons in its outer shell which it shares with its four nearest neighbors. The substitution of an atom from Group V for silicon in the crystal (for example phosphorus) produces a phosphorus atom that shares four of its five electrons (one with each of its four nearest neighbors; see Figure 1–5a). The fifth or extra electron is not needed for bonding purposes and is free to conduct electrical current. Semiconductor regions containing excess conduction electrons are called *n*-type. In an analogous manner, additional holes are provided by substituting an atom like boron for a silicon atom (Figure 1–5b). Semiconductor regions containing excess holes are called *p*-type.

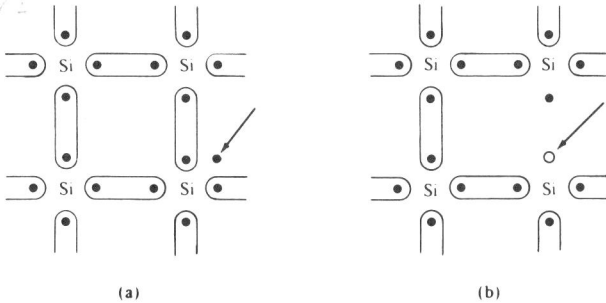

Figure 1–5: The behavior of acceptor and donor atoms in silicon. (a) Extra hole easy to remove; (b) Extra electron easy to remove.

Atoms supplying additional electrons for the conduction process are called *donors;* the number of donors/cm^3 in a semiconductor is N_D. Atoms that supply additional holes for the conduction process are called *acceptors;* the number of acceptors/cm^3 in a semiconductor is N_A. In silicon, potential donor atoms are the atoms of Group V, containing five electrons in their outer shell. The atoms frequently used to dope silicon *n*-type are phosphorus, arsenic, and antimony. Potential acceptors for silicon are the Group III atoms, containing three electrons in their outer shell. The atoms used to dope silicon *p*-type are boron, aluminum, and gallium. (Boron is used most frequently.)

An increase in the number of conduction electrons present in a semiconductor causes a corresponding decrease in the number of holes, and vice versa. The equation $n \cdot p = n_i^2$ is valid even when n does not equal $p(n \neq p)$. If only donor atoms have been added to silicon, and the number of donors is less than 10^{19}/cm^3 ($N_D < 10^{19}$/cm^3), all of the donors produce conduction electrons. It follows that, in this case,

$$n = N_D$$

and

$$p = \frac{n_i^2}{N_D}$$

In a similar fashion, if only acceptors are added to a bar of silicon, and their number is less than 10^{19}/cm^3 ($N_A < 10^{19}$/cm^3), each acceptor atom produces one hole. In this case,

$$p = N_A$$

and

$$n = \frac{n_i^2}{N_A}$$

When both donors and acceptors are added to a semiconductor, they tend to cancel each other out. When more donor than acceptor atoms are added ($N_A < N_D$), the donor atoms cancel out the effect of all of the acceptor atoms, and the number of electrons is the difference between the number of donors and the number of acceptors ($n = N_D - N_A$). In an analogous manner, if more acceptor atoms are added than donor atoms ($N_D < N_A$), the acceptor atoms cancel out the effect of all of the donor atoms, and the number of holes is the difference between the number of acceptors and the number of donors ($p = N_A - N_D$). In both cases, the product $n \cdot p$ remains constant, so the carrier type in the minority are determined using the formula $n \cdot p = n_i^2$.

1.4 Resistivity of a Semiconductor Doped with a Single Impurity

The amount of dopant present in a semiconductor is determined by measuring its conductivity or resistivity. The resistivity of a material is the opposing force a material exerts to prevent flow of current when a voltage is placed across it. The symbol for resistivity is the Greek letter ρ. The units of resistivity are ohm-centimeters (Ω-cm). Conductivity is related to resistivity by the equation

$$\sigma = \frac{1}{\rho} \tag{1-3}$$

The conductivity of a sample depends upon the number of free carriers (holes and/or conduction electrons) and their mobility—the ease with which they move through the sample. If the resistivity (or the conductivity) of a material is known, the resistance of a rectangular block of the material is determined by the formula

$$R = \frac{\rho l}{w \times t} = \frac{\rho l}{A} \tag{1-4}$$

where

R = the resistance of the material (units of ohms)
l = the length of the material from contact to contact
w = width
t = thickness
A = the cross-sectional area (area = thickness $\times$ width).

Figure 1–6: The resistance of a rectangular block of material.

The resistance of a piece of material is related to the applied voltage (V) and the current that flows (I) by the equation

$$V = RI \qquad (1\text{-}5)$$

or, equivalently,

$$R = \frac{V}{I}$$

In the semiconductor (and in other industrial materials as well), the "sheet resistance" of a material is an often-measured parameter. The symbol for sheet resistance is R_s. Sheet resistance is measured in ohms per square ($\Omega/\square$). The resistance of a resistor made up of n squares laid in a row is nR_s. (For instance, if 10 squares of material are laid in a row with $R_s = 100\ \Omega/\square$, $nR_s = 10R_s = 1000\ \Omega$.) Sheet resistance is measured using a four-point probe (Figure 1-7). The formula relating sheet resistance to current and voltage is

$$R_s = 4.53 \frac{V}{I} \qquad (1\text{-}6)$$

This equation is valid when:

1. The thickness of the layer being measured is much less than the spacing between the probes, and
2. The size of the piece of material being measured is much greater in length and width than the probe spacing.

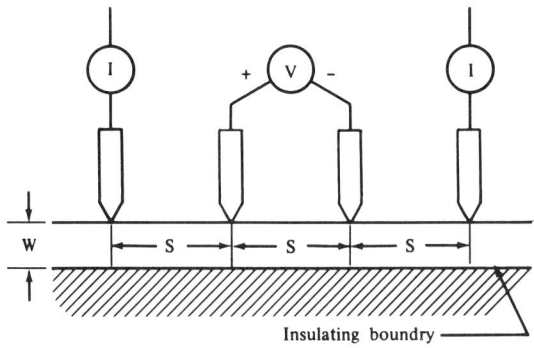

Figure 1-7: Four-point probe.

1.4 Resistivity of a Semiconductor Doped with a Single Impurity

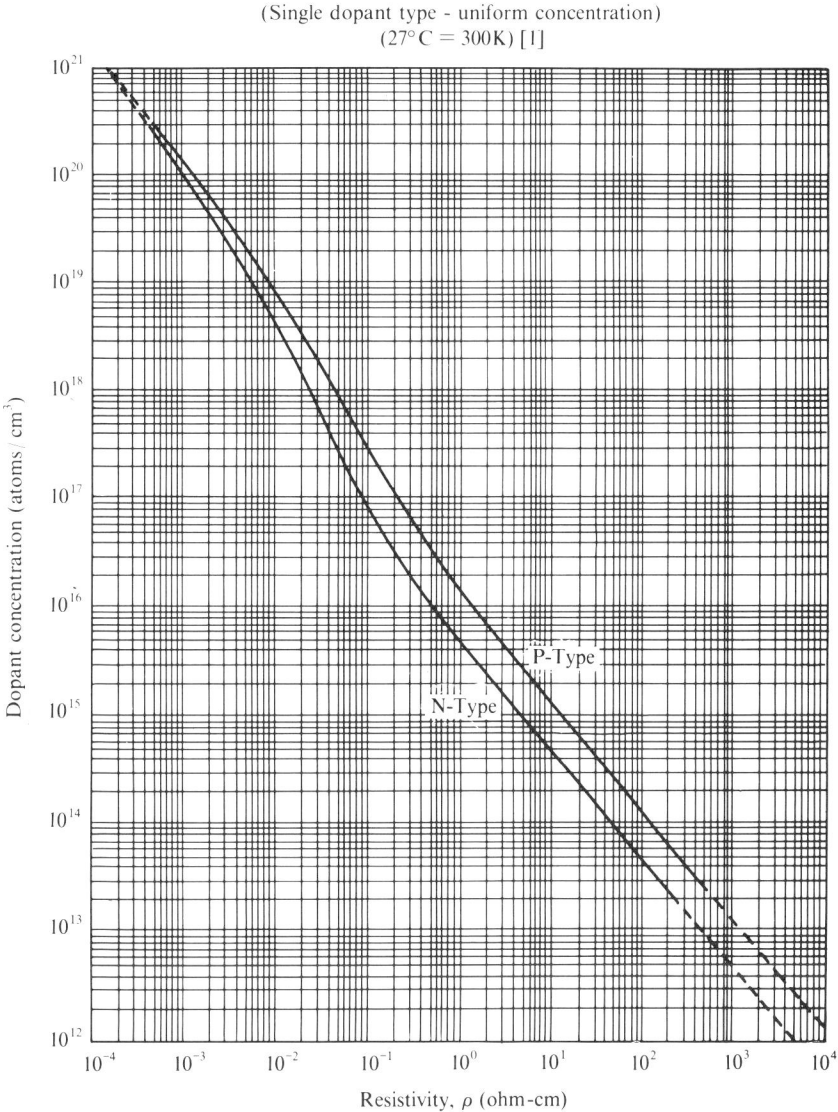

Figure 1–8: Resistivity vs. impurity concentration for silicon with just n-type or p-type dopants.

If a thin layer of material is uniformly doped and the sheet resistance is known, the resistivity ρ is found by using the equation:

$$\rho = R_s \times \text{thickness}, \tag{1-7}$$

or

$$\rho = R_s \cdot t$$

If the thickness of the sample is much greater than the probe spacing, the formula

$$\rho = 2\pi S \frac{V}{I} \quad (\pi = 3.14159) \tag{1-8}$$

relates the current and voltage readings of a four-point probe to the resistivity of the material, where S is the spacing between the probes.

The resistivity of silicon depends upon both the number of acceptor and donor atoms added and the temperature. When just acceptor or just donor atoms are added to a bar of silicon, the resistivity of the silicon can be obtained from Figure 1-8. Conversely, the doping concentration of a uniformly doped sample may be determined if the resistivity is known. To determine the resistivity of a sample when the doping concentration and type are known, find the doping concentration along the bottom of the graph, then proceed upward until the line corresponding to the types of dopant (*p*-type or *n*-type) is encountered. Then proceed horizontally to the left to obtain the resistivity. The reverse operation is performed to obtain the doping concentration from the resistivity.

1.5 Resistivity of a Semiconductor Doped with Both *n*-type and *p*-type Impurities

The addition of impurities to a semiconductor modifies the resistivity in a manner that produces useful behavior in the semiconductor. The conductivity of a material depends on the number of holes and conduction electrons there are in it, the charge each particle carries (called q, the charge of an electron $= 1.6 \times 10^{-19}$ coulombs), and the ease with which the holes and electrons move through the material. The formula for the conductivity of a material can be written as

$$\sigma = qn\mu_n + qp\mu_p \tag{1-9}$$

where

μ_p = hole mobility

μ_n = electron mobility

1.5 Resistivity of a Semiconductor Doped with Both n-type and p-type Impurities

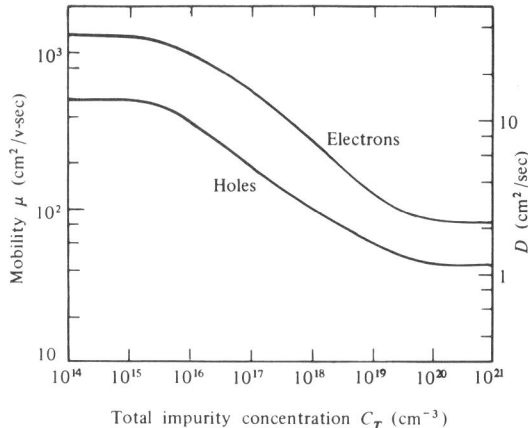

Figure 1–9: Electron and hole mobilities in silicon.

The terms n and p are determined as previously discussed. The ease with which carriers (either holes or electrons) move through a crystal is influenced by the total number of impurity atoms present. Each impurity atom in the lattice is a slight disruption in the otherwise regular crystal structure. A graph of the mobility of both holes and electrons in silicon at 27°C is shown in Figure 1–9. The total dopant concentration C_T is the sum of the number of donor and acceptor atoms:

$$C_T = N_A + N_D \qquad (1\text{--}10)$$

The graph of resistivity versus impurity concentration shown in Figure 1–8 is valid when only acceptor or only donor atoms are added to silicon. If both types of impurities are added, the resistivity of the material must be calculated.

EXAMPLE 1

$N_D = 2 \times 10^{15}/\text{cm}^3$; $N_A = 4 \times 10^{15}/\text{cm}^3$ at 27°C; determine the resistivity of the sample.

First, determine n and p.
Since N_A is greater than N_D,
$p = N_A - N_D = 2 \times 10^{15}/\text{cm}^3$

$n = \dfrac{n_i^2}{p} = 1 \times 10^5$

μ_n and μ_p are found by noting that $N_D + N_A = 6 \times 10^{15}/\text{cm}^3$

$\mu_n = 1100 \text{ cm}^2/\text{V} - \text{sec}; \mu_p = 400 \text{ cm}^2/\text{V} - \text{sec}$

$\sigma = q(\mu_n n + \mu_p p) \cong q\mu_p p$

Now

$$\rho = \frac{1}{\sigma} = \frac{1}{(1.6 \times 10^{-19})(400)(2 \times 10^{15})} = 7.8 \text{ }\Omega\text{-cm}$$

EXAMPLE 2

$N_D = 6 \times 10^{17}/\text{cm}^3$; $N_A = 3 \times 10^{17}/\text{cm}^3$ at 27°C; determine the resistivity of the sample.

First, determine n and p.
Since N_D is greater than N_A,

$n = N_D - N_A = 3 \times 10^{17}/\text{cm}^3$

$p = \dfrac{n_i^2}{n} = 6.7 \times 10^2/\text{cm}^3$

μ_n and μ_p are found by first determining C_T
$= N_A + N_D$; thus $\mu_n \cong 700$, $\mu_p \cong 200$.

$\sigma = q(\mu_n n + \mu_p p) \cong q\mu_n n$

$ = (1.6 \times 10^{-19})(700)(3 \times 10^{17})$

$\sigma = 3.36 \times 10^1 = 33.6 \dfrac{1}{\Omega\text{-cm}}$

$\rho = 0.28 \text{ }\Omega\text{-cm}$

1.6 | Carrier Transport

The carriers present in a semiconductor move by one of two processes, drift or diffusion. *Drift* is the motion caused by the presence of an electric field. With no applied field, carriers move about randomly in a semiconductor (Figure 1–10a). Under the influence of an applied field, the carriers acquired a directed component of motion (Figure 1–10b). The sum of the directed components of drift produce current flow in the sample. *Diffusion* is the migration of particles from regions of high concentration to regions of low concentration caused by random motion. The resultant magnitude and direction of both types of carrier motion determines the total current flow in a material.

The diffusion of the mobile carrier (either holes in *p*-type silicon or conduction electrons in *n*-type silicon) from higher temperature regions to lower temperature regions is used to determine whether a sample of silicon is *n*-type or *p*-type. If an area of a silicon wafer is heated locally, as shown in Figure 1–11, the majority carriers diffuse away from the heated region. A voltage results

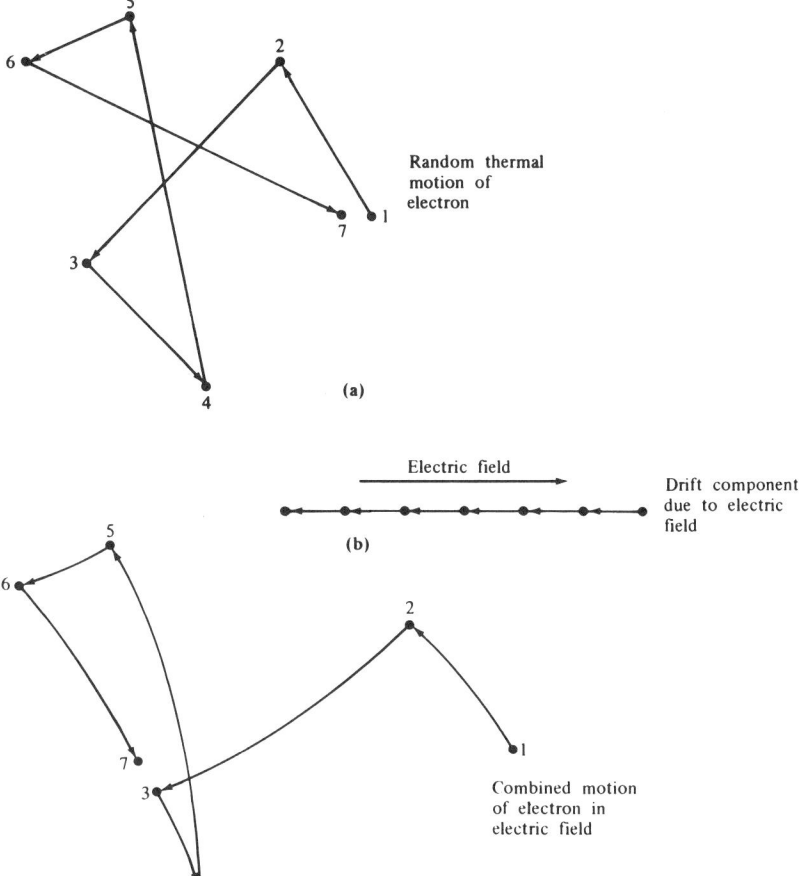

Figure 1–10: Carrier drift under the influence of an electric field.

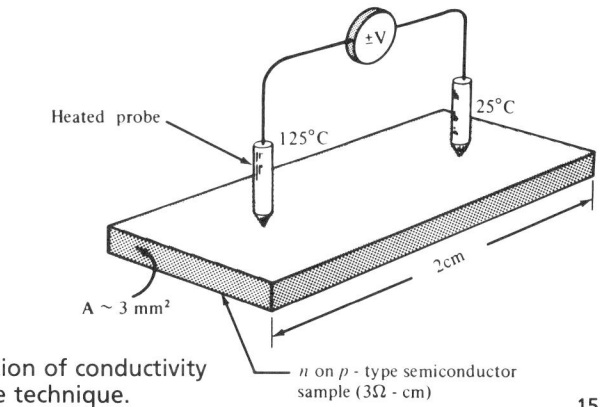

Figure 1–11: Determination of conductivity type using the hot-probe technique.

which is measured to determine the conductivity of the sample. If the sample is *n*-type, the voltage on the hot probe is positive with respect to a second probe. Similarly, if the sample is *p*-type, the voltage on the hot probe is negative with respect to a second probe. This testing technique is useful on samples unless there is just a thin layer of oppositely doped material on the surface of the wafer.

Review Exercises

1. A sample of silicon is doped with 10^{15} atoms/cm^3 of phosphorus.
   a. Determine the donor concentration N_D.
   b. Determine the acceptor concentration N_A.
   c. Determine the electron concentration n.
   d. Determine the hole concentration p.
   e. Determine the resistivity ρ.

2. A sample of silicon is doped with 2×10^{16} atoms/cm^3 of boron.
   a. Determine the donor concentration N_D.
   b. Determine the acceptor concentration N_A.
   c. Determine the electron concentration n.
   d. Determine the hole concentration p.
   e. Determine the resistivity ρ.

3. If a sample of silicon is doped with 3×10^{17} atoms/cm^3 of arsenic and 5×10^{17} atoms/cm^3 of boron,
   a. Determine the donor concentration N_D.
   b. Determine the acceptor concentration N_A.
   c. Determine the electrons concentration n.
   d. Determine the hole concentration p.

4. A four-point probe measurement has been made on a sample yielding the following results:

$$V = 5 \times 10^{-3} \text{ volts}$$
$$I = 4.5 \times 10^{-3} \text{ amps}$$

What is the sheet resistance, R_S, of the sample?

5. A sample of material has the following properties:

   Its length is 100 micrometers.
   Its width is 5 micrometers.
   Its thickness is 2 micrometers.
   Its resistivity is 2 Ω-cm.

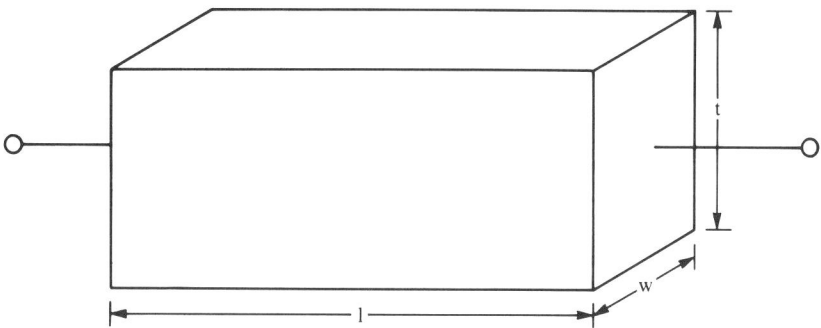

Determine the resistance of the bar of material.

6. A silicon sample is doped with 2×10^{16} acceptors/cm^3 and 5×10^{15} donors/cm^3. What type of impurity in what concentration should be added to make the equilibrium electron and hole concentrations the same at room temperature?

7. A silicon bar with a length of 1 cm and a thickness and width of 0.1 cm has a resistance measured from end to end of 10 Ω. If a hot probe measurement indicates the bar to be *n*-type, determine the donor concentration.

8. A bar of silicon is doped with 2×10^{15} arsenic atoms/cm^3. Calculate the resistivity of this bar and compare your answer with Figure 1–8.

9. A bar of silicon contains 1×10^{18} boron atoms/cm^3 and 3×10^{18} antimony atoms/cm^3.
    a. Determine the donor and acceptor concentrations, N_D and N_A.
    b. Determine the hole and electron concentrations, p and n.
    c. Determine the mobilities, μ_p and μ_n, of the holes and the electrons.
    d. Determine the resistivity of the bar.
    e. Why does the answer to part d differ from that obtained from Figure 1–8 with $N_D = 3 \times 10^{18}$/cm^3.

10. A piece of silicon is doped with 7×10^{15} boron atoms/cm^3 and 3×10^{15} phosphorus atoms/cm^3. Find the electron and hole concentrations at 27°C.

2 | Crystal growth and wafer preparation

2.0 | Introduction

The silicon used in the fabrication of semiconductor devices is extremely pure. Before device manufacture is started, the silicon has a typical impurity concentration of less than one part per billion. The starting point for virtually all semiconductor device technology is this silicon in the form of flat, circularly shaped substrates called *wafers*. The atoms of silicon within these wafers are arranged in a repetitive fashion on a long-range basis, and the resulting material is said to be *crystalline*. The steps required to convert randomly ordered or polycrystalline material (material containing many crystals) into silicon wafers having long-range order are a complex procedure that is a story in itself.

Silicon is one of the most abundant materials on the earth's surface, representing approximately 25% of the earth's crust. However, silicon combines quite easily with oxygen and other materials to form compounds. The silicon must first be separated from these compounds and extensively refined before it can be transformed into the single-crystal wafers used to manufacture semiconductor devices. Sand, in the form of silicon dioxide (SiO_2), is found in many locations with impurity levels of less than 1%. This sand is used as the starting point for the manufacture of silicon wafers. Once it is refined into polycrystalline silicon, large crystals are formed which are then sliced into wafers. The next section discusses the steps necessary for the production of ultrapure silicon for semiconductor devices.

2.1 Origin of Silicon and Its Purification

There are three basic steps which are used in the preparation of silicon for use in wafer manufacturing: the production of the polycrystalline silicon, crystal growth, and the actual manufacture of the wafer. The following steps are practiced in the production of ultrapure polycrystalline silicon for the manufacture of silicon wafers to be used for semiconductor devices.

STEP 1

Two of the most common materials on earth, silicon dioxide (sand) and carbon, are made to react with each other at high temperatures in an electric arc furnace to form silicon (99% pure) and carbon monoxide. The silicon is formed when the silicon dioxide (SiO_2) is reduced by the carbon. The reaction is

$$SiO_2 + 2C \xrightarrow{heat} Si \uparrow + 2CO \uparrow \qquad (2\text{--}1)$$

This reaction produces silicon in vapor form at the reaction temperature and a gaseous by-product, which are both easily exhausted. The silicon vapor is then condensed to give a product known as metallurgical grade silicon which is 99% pure—far from the quality of silicon required for modern-day semiconductor device technology. Thus, further steps are necessary to remove additional undesirable impurities.

STEP 2

The metallurgical grade silicon is further purified by converting it to trichlorosilane ($SiHCl_3$). Trichlorosilane is a liquid at room temperature and is easily purified to semiconductor standards by fractionation (distillation) procedures. Once the trichlorosilane has been purified, it is reduced in hydrogen to form polycrystalline silicon. The reaction to form trichlorosilane is

$$Si + 3HCl \xrightarrow{1,260°C} SiHCl_3 + H_2 \uparrow \qquad (2\text{--}2)$$

The fractionation process separates the trichlorosilane from silicon tetrachloride ($SiCl_4$) and other chlorides of dopant impurities such as phosphorus and boron, and chlorides of metal such as iron and copper.

2.1 Origin of Silicon and Its Purification

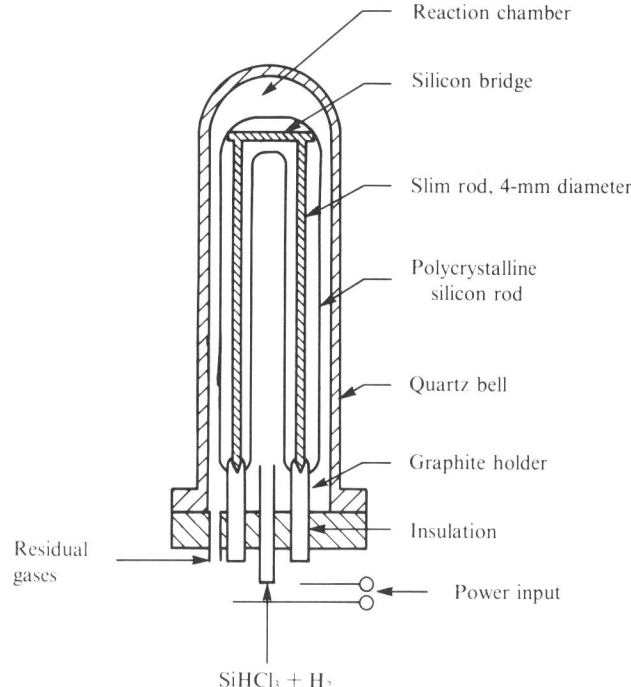

Figure 2–1: Schematic of a CVD reactor used for electronic grade silicon production. *(After Crossman and Baker, Ref. 1. This figure was originally presented at the spring 1977 meeting of the Electrochemical Society, Inc. held in Philadelphia, Pennsylvania.)*

STEP 3

Polysilicon deposition is accomplished by the thermal reduction of pure trichlorosilane. The reaction takes place in a reactor where the gas flows past pure silicon rods called "slim rods" which are resistively heated. The rods serve as a nucleation point for the deposition of silicon. The reaction is

$$SiHCl_3 + H_2 \xrightarrow{heat} Si + 3HCl \qquad (2-3)$$

The apparatus for such a process is illustrated in Figure 2–1.

The three steps take several hours and produce polysilicon rods up to 20 cm in diameter and 2–3 meters in length. The polycrystalline silicon is now ready for the crystal-growing process.

2.2 | Crystal Growth

There are presently two methods used to grow single-crystal silicon for semiconductor applications: the *Czochralski* and *float-zone* crystal growth techniques (abbreviated CZ and FZ, respectively). The CZ method is by far the most popular method, accounting for between 80 and 90% of all the silicon crystals grown for the semiconductor industry. With the exception of those processes which require extremely low oxygen content in the wafer, virtually all of the silicon produced for integrated circuit fabrication is prepared by the CZ technique.

CZ crystal growth utilizes a quartz (SiO_2) crucible of high purity in which pieces of polycrystalline silicon (termed "charge") are heated to their melting

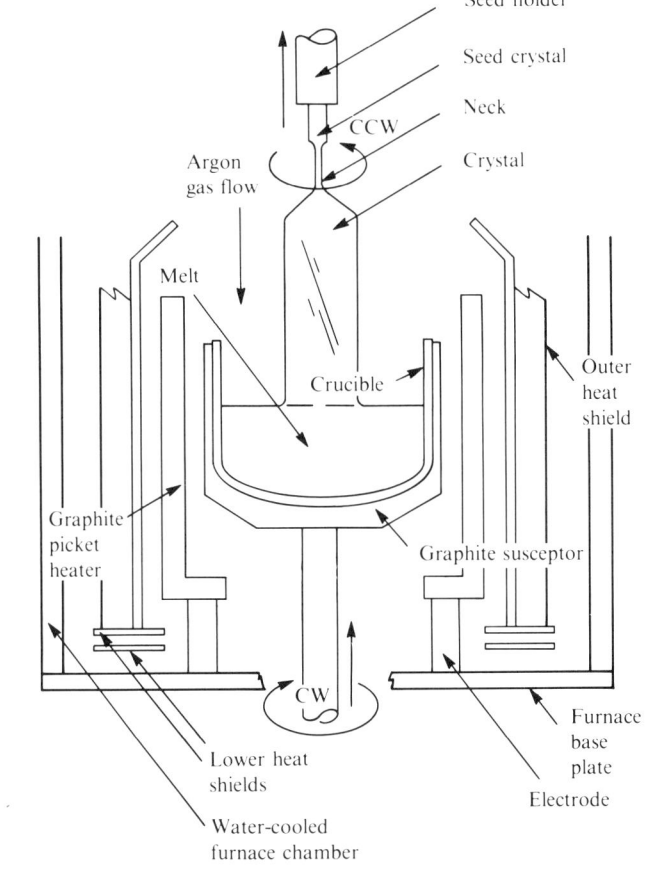

Figure 2–2: Cross section of a furnace for growing single-crystal silicon by the Czochralski process. *(After Bonora, Ref. 2.)*

point of 1415°C. The crucible, shown in Figure 2–2, is heated by either induction using radio-frequency (RF) energy or thermal resistance methods. A seed crystal with the desired orientation is dipped into the molten silicon (termed "melt") and then withdrawn at a carefully controlled rate. When the procedure is properly done, the material in the melt will make the transition to a solid-phase crystal (that is, it will "freeze") at the solid–liquid interface, so the newly created material accurately replicates the crystal structure of the "seed" crystal. The pull rate of the seed crystal varies during the growth cycle. It is fastest when growing the relatively narrow "neck," (5–12 inches per hour) so the generation of defects known as dislocations is minimized. Once the neck has been formed, the pull rate is reduced to form the shoulder of the crystal, finally approaching 2–4 inches per hour during the growth of the crystal body.

During the entire growth process, the crucible rotates in one direction at 12–14 rpm while the seed holder rotates in the opposite direction at 6–8 rpm. This constant stirring prevents the formation of local hot or cold regions. Crystal diameter is monitored by an optical pyrometer which is focused at the interface between the edge of the crystal and the melt. An automatic diameter control (ADC) system maintains the correct crystal diameter by closed-loop control. Argon is often used as the ambient gas during this crystal-pulling process. By carefully controlling the pull rate, the temperature of the crucible, and the rotation speed of both the crucible and the rod holding the seed, precise control of the diameter of the crystal is obtained.

The desired impurity concentration is obtained by adding impurities to the melt in the form of small amounts of heavily doped silicon prior to crystal growth.

As the crystal is grown, the melt level in the crucible drops relative to the hot-zone area of the furnace. To maintain the crystal-melt interface within the hot zone of the furnace, an electromechanical crucible lift system elevates the crucible continuously during growth.

FZ crystal growth proceeds directly from the rod of polysilicon obtained in the silicon purification process. Thus, the method uses no crucible. (See Figure 2–3.)

A rod of polycrystalline silicon of the appropriate diameter is held at the top and placed in the crystal-growing chamber. A single crystal seed is clamped in contact with the other end of the polycrystalline silicon rod. The rod and the seed are enclosed in a chamber with a controlled, inert atmosphere, and an induction heating coil is placed around the rod. The coil melts a small length of the rod, starting with part of the single seed crystal. A "floating zone" of melt is formed between the seed crystal and the polysilicon rod. The molten zone is then slowly moved up along the length of the rotating rod by moving the coil upward.

The molten region that solidifies first remains in contact with the seed crystal and assumes the same crystal structure as the seed. As the molten region is moved along the length of the rod, the polycrystalline rod melts and

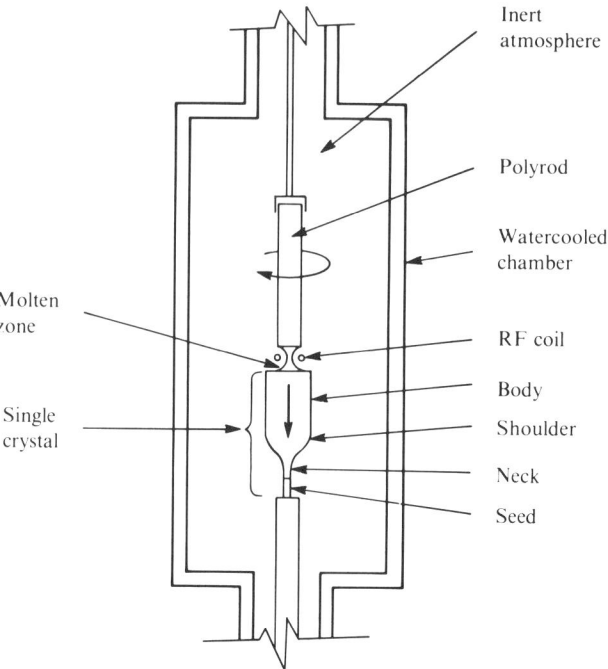

Figure 2–3: Cross section of a furnace for growing single-crystal silicon by the float-zone process.

then solidifies along its entire length, becoming a single crystal rod of silicon. The diameter of the crystal is controlled by the motion of the heating coil. Because of the difficulties in preventing collapse of the molten region, this method has been limited to small-diameter (less than 76 mm) crystals. However, since there is no crucible involved in the FZ method, oxygen contamination that might arise from the quartz (SiO_2) crucible is eliminated. Wafers manufactured by this method find use in applications requiring low-oxygen-content, high-resistivity starting material for devices such as power diodes, power transistors, and SCRs.

2.3 Doping of Crystals During Growth

During the CZ method of crystal growth, dopant is introduced into the crystal by adding small amounts of heavily doped silicon to the melt. Normally, as the silicon freezes, dopant is rejected back into the melt. The concentration of dopant in the crystal is thus less than that in the melt, and as the melt is consumed, it becomes richer in dopant. As a result, the seed end of the crystal is more lightly doped and, hence, has a higher resistivity than the tail end of the crystal. The actual concentration of the dopant along the crystal depends on both the material used as the dopant and its original concentration. A quantity called the *distribution coefficient*, denoted by the letter K, is a measure of the ratio of the concentration of the dopant in the solid to the concentration of the dopant in the liquid:

$$K = \frac{C_s}{C_\ell} = \frac{\text{concentration of dopant in the solid phase}}{\text{concentration of dopant in the liquid phase}} \quad (2\text{-}4)$$

The distribution coefficients of various *n*-type and *p*-type dopants are given in Table 2–1.

Typical axial (end-to-end) resistivity variations for CZ crystals, along with the appropriate distribution coefficients, are shown in Figure 2–4.

The FZ crystal growth method commonly utilizes boron or phosphorus dopant, either of which is introduced as a gas. For this reason, FZ crystal growth, unlike that of CZ method, achieves very good axial resistivity control.

2.4 | Wafer Manufacture

The first step in the manufacture of silicon wafers from a grown crystal is to grind the crystal perfectly round and determine its rotational orientation. The seed crystal has determined which crystal face will be present on the wafer surface, but the rotational position of the rod determines other axes of the crystal. Since the ingot of silicon is one crystal, it has preferential break or cleavage planes. It is critical for later device separation (scribe and break) to align the circuits precisely with respect to the cleavage planes. This precise

Table 2–1: Distribution Coefficients for Common Dopants in Silicon

DOPANT	DISTRIBUTION COEFFICIENT k	TYPE OF DOPANT
Phosphorus (P)	.35	*n*-type
Arsenic (As)	.30	*n*-type
Antimony (Sb)	.023	*n*-type
Boron (B)	.80	*p*-type
Aluminum (Al)	1.8×10^{-3}	*p*-type
Gallium (Ga)	9.2×10^{-3}	*p*-type
Indium (In)	3.6×10^{-4}	*p*-type

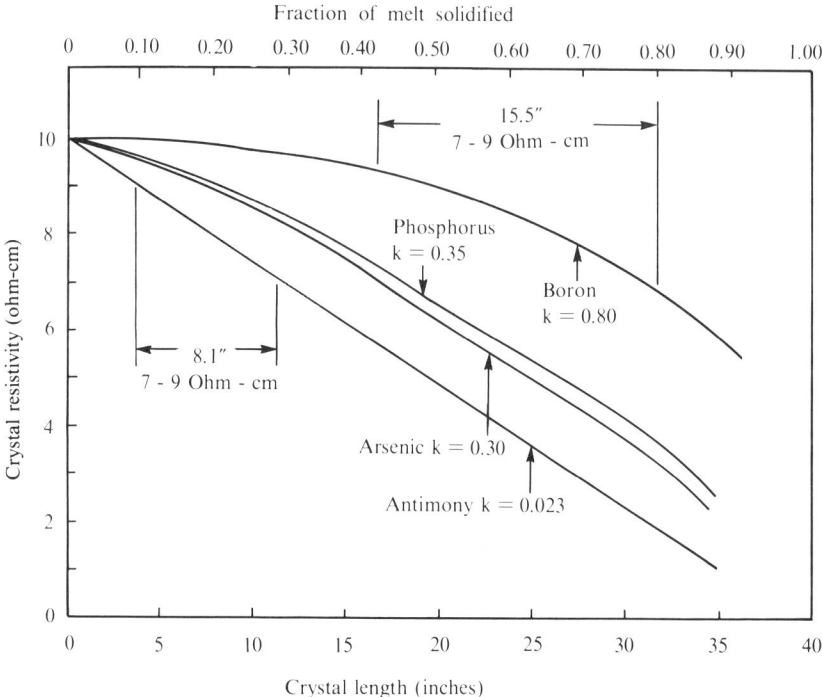

Figure 2–4: Crystal resistivity along a Czochralski crystal. *(After Trapp, Blanchard, and Shepherd, Ref. 3.)*

alignment is accomplished by grinding one or more flats along the crystal prior to sawing the crystal into wafers. The flats are then used as a reference during all subsequent processing steps. X-ray diffraction provides a fast and accurate method for determining the crystal orientation prior to grinding the flats. After grinding, the surface of the crystal is chemically etched to remove any damage that might have been incurred.

The silicon crystal is then sawed (sliced) into thin slices called *wafers*, using the major flat as a reference. Extreme care is taken to minimize the amount of single-crystal silicon lost during the sawing process. The inside diameter of a ring-shaped saw blade performs the cutting. The cutting edge of the blade is coated with diamond powder to enable it to cut through the hard silicon. Problems which may occur during the sawing operation and which must be controlled include the occurrence of chips and edge flakes, exit damage, and bow. A silicon etchant is used to remove saw marks and any accompanying damage from both sides of the wafer after sawing.

Normally, an "as-sawed" wafer is not sufficiently flat to meet polished slice requirements. Lapping, a free-abrasive machining process, is used to planarize both sides of the slice and provide a uniform finish. Silicon material is removed simultaneously from both sides of the slice on a double-sided lapping machine.

Before or after lapping, the wafer is processed through an edge-rounding machine which removes the square-cornered edge, which tends to be laden with microcracks from sawing and is prone to fracture. Other advantages of edge rounding include reduction of contamination from silicon chips, easier loading into processing equipment, reduction of photoresist edge build-up, and improved mask life for proximity and contact printing. Disadvantages include loss of diametral accuracy, rounding of flat cusps, angular shifting of the flat, and scalloping of the wafer edge.

Lapping damage is removed by chemical etching, a damage-free means of removing material from the surface. The two processes currently in use are acid and caustic etching, which remove 10 to 30 micrometers from each side of the wafer. Acid etchants are mixtures of hydrofluoric, acetic, and nitric acids. Because of the hazards involved in acid etching, caustic etching in hot (70 to 130°C) potassium or sodium hydroxide is gaining favor. Although uniformity is improved and environmental pollution is reduced, the etch cycle is considerably longer for caustic etching than for acid etching.

After an intermediate inspection for thickness, flatness, and resistivity, the wafers are moved on to the polishing process. Here, they are mounted on a polisher, and one side of the wafer receives a mirror-like finish. The polishing operation uses a polishing solution that simultaneously chemically etches and mechanically polishes the wafers. The polishing pad must be tough and durable. When the wafers have reached the proper surface quality and range of thickness, the polishing plates are removed and the wafers are dismounted. The wafers are thoroughly cleaned to remove any residual contamination and are inspected for particulate contamination, resistivity, thickness, and flatness. Wafers that pass the final inspection are ready to start on their journey to becoming devices.

2.5 | Crystal Orientation

The orientation of a silicon crystal is an important parameter in the device fabrication sequence. One method used to describe the orientation of crystals is through the use of *Miller indices*. The Miller indices of a plane of silicon atoms are determined by the points at which the crystal plane intersects the x, y, and z-axes illustrated in Figure 2–5. Equations 2–5 through 2–7 show the relationships involved:

$$x\text{-index} = \frac{1}{x \text{ point of intersection}} \quad (2\text{–}5)$$

$$y\text{-index} = \frac{1}{y \text{ point of intersection}} \quad (2\text{–}6)$$

$$z\text{-index} = \frac{1}{z \text{ point of intersection}} \qquad (2-7)$$

As an example, consider an intersecting plane and set of axes in Figure 2–5. Suppose the plane intersects the axes at $x = 1, y = 1$, and $z = 1$. The Miller indices are therefore:

$$x\text{-index} = \frac{1}{x \text{ point of intersection}} = \frac{1}{1} = 1$$

$$y\text{-index} = \frac{1}{y \text{ point of intersection}} = \frac{1}{1} = 1$$

$$z\text{-index} = \frac{1}{z \text{ point of intersection}} = \frac{1}{1} = 1$$

The plane is thus the <111> plane of the crystal. Other planes parallel to it are also <111> planes. The <111> planes intersect the crystal axes so that a triangle is formed by their intersection. Accordingly, crystals that are sliced in the <111> plane can be recognized by the triangular pits that will be etched in their surface or the triangular pieces of silicon that result when the wafers are broken.

Consider next the same set of indices, and take a plane that intersects the x-axis at $x = 1$ but never intersects the y- or z-axes, as shown in Figure 2–6. The Miller indices of the plane are:

$$x\text{-index} = \frac{1}{x \text{ point of intersection}} = \frac{1}{1} = 1$$

$$y\text{-index} = \frac{1}{y \text{ point of intersection}} = \frac{1}{\infty} = 0$$

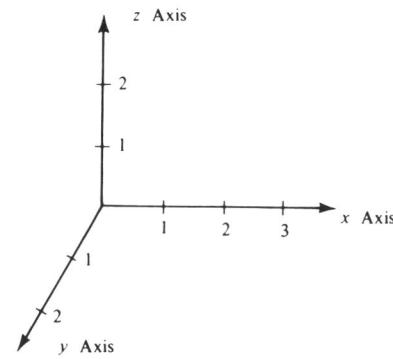

Figure 2–5: The crystal axis set.

2.6 Crystal Defects

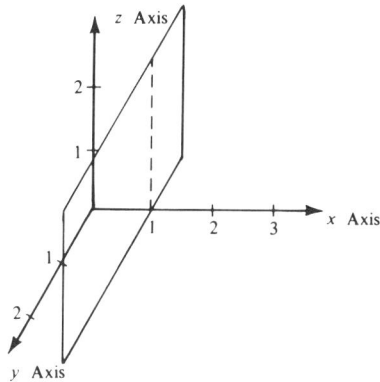

Figure 2–6: A <100> crystal plane.

$$z\text{-index} = \frac{1}{z \text{ point of intersection}} = \frac{1}{\infty} = 0$$

(For mathematical reasons, a plane that does not intersect an axis is considered to intersect the axis at infinity (∞).)

The Miller indices indicate that the plane in question is a <100> plane. The <100> planes intersect the crystal axes so that rectangles are formed. If such a wafer is etched, the resulting etch pits will form with square corners, and the wafer will break into rectangular pieces if it is shattered. The <111> and <100> orientations are the two primary wafer orientations in commercial use.

All wafers are provided with a primary flat so that the device array can be aligned with respect to the scribe and break directions of the wafer. Secondary flats may be added to further identify the doping type, as well as the orientation. The locations of the primary and secondary flats with regard to orientation and type are illustrated in Figure 2–7.

2.6 | Crystal Defects

A variety of crystal defects can be present in grown crystals of silicon and other semiconductor materials. These defects may affect whether wafers can be successfully processed from such crystals or, even if they can, whether adequate yield can be obtained on devices made with these wafers. Two common types of defects frequently encountered are:

1. *Crystal dislocations*—localized imperfections in the crystal structure caused by plastic deformation from uneven heating or cooling or other problems.

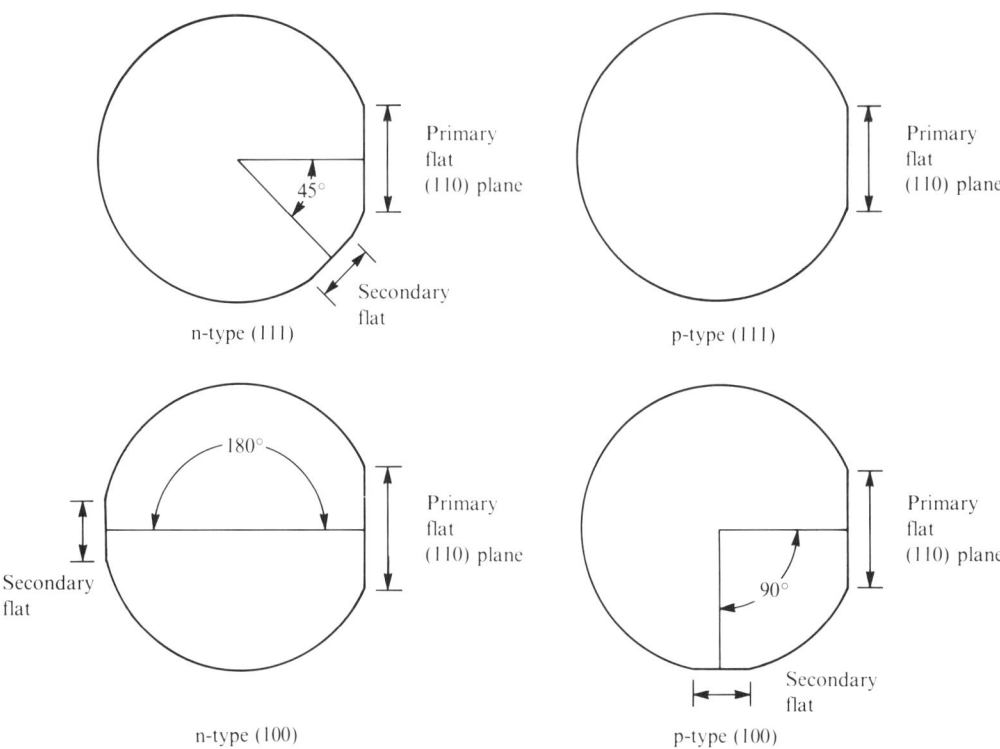

Figure 2–7: Primary and secondary flat locations on silicon wafers.

2. *Planar slip*—a type of plastic deformation visible because one part of the crystal sheared with respect to another.

Both of these common defects, as well as other less common problems, can be decorated using a preferential etch. Such an etch attacks the silicon along defect boundaries, revealing the nature and extent of the defect.

Crystal defects have been virtually eliminated from silicon grown using either the FZ or CZ method. However, defects are often inadvertently introduced during subsequent processing, so their study is of continuing importance. Dislocations, slip, and other defects are usually introduced if wafers are improperly heated or cooled during any one of the high-temperature processing steps.

Review Exercises

1. Why is it desirable to end up with a solid and a gas—or a liquid and a gas—following a chemical reaction?
2. a. What contaminant may CZ silicon contain that FZ silicon does not usually contain?
   b. Where does this contaminant come from?
3. Does silicon or silicon dioxide have a higher melting point? Why?
4. a. Why is the crystal orientation of a wafer important?
   b. How is the orientation denoted?
5. Define the term *polysilicon*.
6. What is the purpose of the argon gas during crystal growth?
7. Why is a seed crystal used for crystal growth?
8. What two variables are used to control the diameter of the silicon rod?
9. List six advantages gained from the edge-rounding step used during wafer manufacture.
10. List eight quality control problems that can arise after the sawing operation during wafer manufacture.
11. a. Determine the Miller indices of a plane intersecting a set of axes at $x = 1/2, y = 1, z = \infty$.
    b. Sketch the plane in the axes provided.

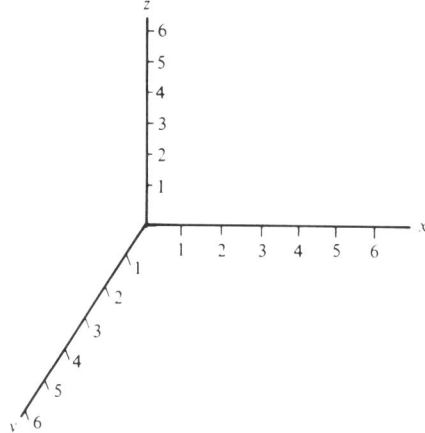

12. What value of K will result in a flat profile for the dopant?
13. What p-type dopant of Table 2–1 will yield the flattest impurity profile during CZ crystal growth?

14. What are the two most common crystal orientations utilized for silicon processing?
15. Describe two types of crystal defects found in wafers cut from grown ingots of silicon.

REFERENCES (CHAPTER 2)

1. L. D. Crossman and J. A. Baker, "Polysilicon Technology," *Semiconductor Silicon,* Electrochem. Soc., 1977, p. 18.
2. A. C. Bonora, "Review of Silicon Manufacturing," Siltec, 1982, p. 1.
3. O. D. Trapp, R. A. Blanchard, and W. H. Shepherd, *Semiconductor Technology Handbook,* Technology Associates, Portola Valley, CA, 1980, pp. 2–7.

3 | Oxidation of silicon

3.0 | Introduction

The ability to grow a chemically stable protective layer of silicon dioxide (SiO_2) on a silicon wafer makes silicon the most widely used semiconductor substrate. The silicon dioxide layer is both an insulating layer on the silicon surface and a preferential masking layer during the fabrication sequence. This chapter covers physical processes that control the oxidation of silicon, the oxide thickness resulting from a given growth cycle, and details on the growth and evaluation of oxide layers.

3.1 | The Growth of Silicon Dioxide

A silicon dioxide layer is grown in an atmosphere containing either oxygen (O_2) or water vapor (H_2O) at temperatures in the range of 900 to 1300°C. The process of oxidation is understood by considering a surface of silicon with an already existing layer of silicon dioxide on it (Figure 3–1).

Except for the first few moments, a silicon slice will have a layer of SiO_2 on it, so this assumption is a valid one. Oxidation takes place when either oxygen or water vapor reacts with the silicon, as shown in the chemical equations

$$Si + O_2 \rightarrow SiO_2 \tag{3-1}$$

and

$$Si + 2H_2O \rightarrow SiO_2 + 2H_2 \uparrow \tag{3-2}$$

3: Oxidation of Silicon

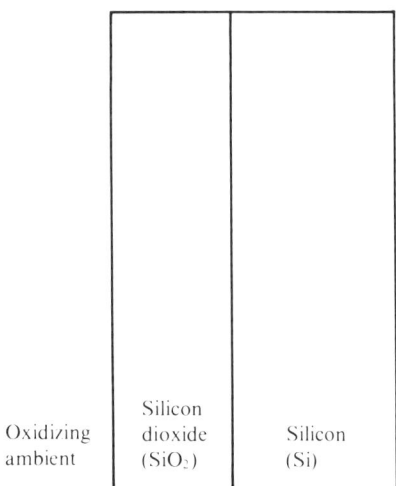

Figure 3–1: Silicon with a layer of silicon dioxide on its surface.

For the silicon and the oxidizing species to react, one of the following must take place:

1. The oxidizing species diffuses through the layer of SiO_2 to reach the silicon–SiO_2 interface, where the reaction takes place (Fig 3–2a).

2. The silicon diffuses through the layer of SiO_2 to its surface, where the reaction takes place (Figure 3–2b).

3. The two active species meet somewhere in the SiO_2 layer where the reaction takes place (Figure 3–2c).

Experiments have disclosed that in the thermal oxidation of silicon, the first process, diffusion of the oxidizing species (either O_2 or H_2O) through the existing layer of SiO_2, occurs.

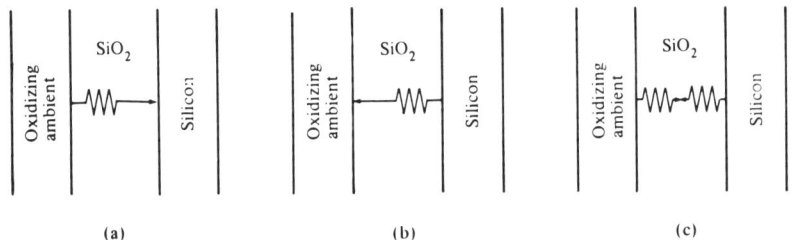

Figure 3–2: Potential reaction mechanisms in the oxidation of silicon. (a) Oxidizing species diffuses through the SiO_2 layer interface; (b) Silicon diffuses through the SiO_2 surface; (c) Oxidizing species and the silicon meet in the SiO_2 layer.

3.2 Equipment for Thermal Oxidation

Thermal oxidation is performed in furnaces where the temperature is carefully controlled. (Limits of $\pm \frac{1}{2}°C$ are typical.) A furnace stack or furnace bank generally contains three or four separate furnaces, each with its own set of controls and quartzware. The furnace heating elements, or coils, are heated by electrical current. The amount of electrical current is controlled and adjusted to provide the required constant temperature. A quartz tube (or occasionally a tube made from another material such as polycrystalline silicon or silicon carbide) rests inside the coils, providing an enclosure around the wafers in which the atmosphere can be controlled. A cross section of a typical oxidation furnace is shown in Figure 3–3.

3.3 The Oxidation Process

Thermal oxidation of silicon is preceded by a cleaning sequence designed to remove all contamination. Special care must be taken during this step to guarantee that the wafers do not contact any source of contamination—particularly inadvertent contact with a person. (Humans are a potential source of sodium, the element most often responsible for the failure of devices due to surface leakage.) The cleaned wafers are dried and loaded into a quartz water holder called a "boat," after which they are ready for oxidation.

Thermal oxidation using dry oxygen involves controlling the flow of oxygen into the quartz tube to guarantee that an excess of oxygen is available for the silicon. A source of high-purity oxygen makes sure that no unwanted impurities are incorporated in the layer of oxide as it grows. Oxygen or an oxygen–nitrogen mixture is used for growing the layer of oxide. The use of nitrogen decreases the total cost of running the oxidation process, since it is less expensive than oxygen.

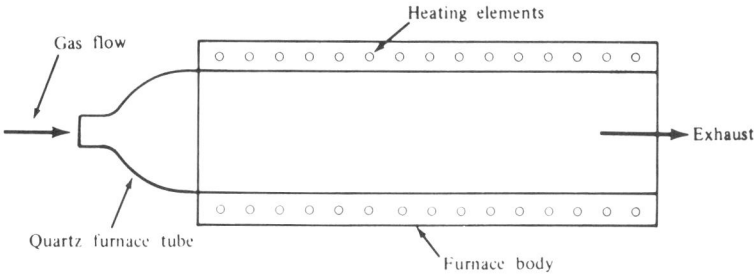

Figure 3–3: Cross section of an oxidation furnace.

Two methods of introducing water vapor are commonly used when water is the oxidizing species. In the first of these, water is placed in a container called a "bubbler" and maintained at a constant temperature below its boiling point (100°C). Figure 3–4 shows a bubbler with a heating mantle that keeps it at a predetermined temperature.

Gas enters the inlet side of the bubbler, becomes saturated with water vapor as it rises through the water, and exits through the outlet into the furnace. The distance from the outlet to the quartz oxidation tube must be short enough to prevent condensation by cooling. Nitrogen or oxygen may be used as the carrier gas, with equivalent oxide thickness being grown regardless of which gas is used. Maintenance of a constant temperature is important because the vapor pressure of water varies with temperature, as shown in Figure 3–5.

The temperature of the water is kept a couple of degrees below its boiling point to afford better control of the vapor pressure. If the temperature is too close to 100°C, small temperature variations produce large changes in the vapor pressure. Bubblers are simple to use and quite reproducible, but they have two disadvantages associated with the fact that they must be refilled when the water level falls too low:

1. Improper handling of the container can result in contamination of the water prior to or during filling.

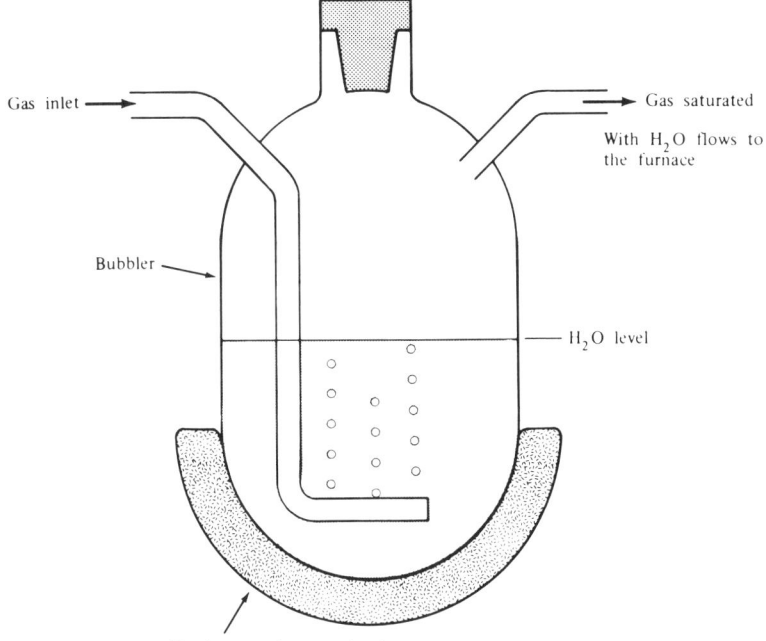

Figure 3–4: A bubbler for a wet oxidation system.

3.3 The Oxidation Process

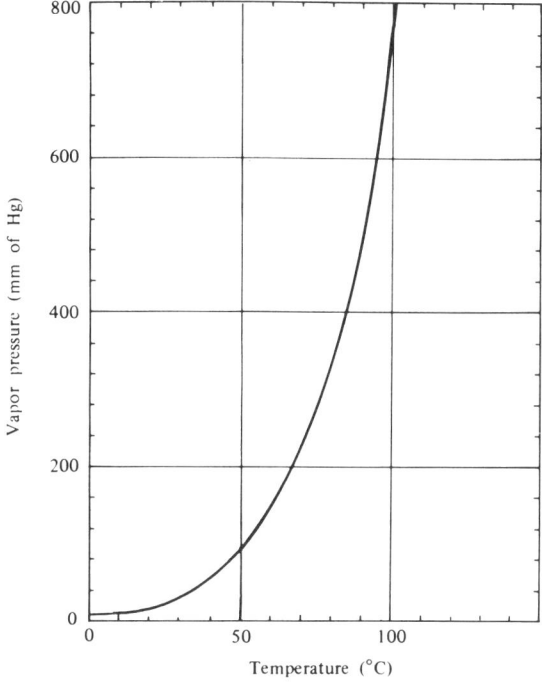

Figure 3–5: The vapor pressure of water as a function of temperature.

2. The bubbler cannot be filled during a cycle unless heated water is used. If cool water is added, the vapor pressure of the water decreases when the new water is introduced.

A second method of obtaining water is the introduction and subsequent combustion of a hydrogen–oxygen gas mixture. Such systems are often called "burnt hydrogen" or "torch" systems. A more accurate term is pyrogenic steam. Water vapor is produced when proper amounts of hydrogen and oxygen are introduced into the inlet end of the tube and allowed to react. A quartz injector with a specially shaped tip is used to guarantee proper combustion of the mixture. The heat produced at the inlet end of the oxidation furnace often makes it necessary to reprofile the furnace under operating conditions in order to guarantee a uniform temperature profile along the entire tube. The installation of these units necessitates the availability of both oxygen and pure hydrogen. Because the potential for an explosion exists if excess hydrogen is introduced, an explosion prevention system is present on installed units.

Table 3–1: Color Chart for Thermally Grown SiO₂ Films
(OBSERVED PERPENDICULARLY UNDER DAYLIGHT FLUORESCENT LIGHTING)

micro → Å (multiply 1000)

FILM THICKNESS (MICROMETERS)	ORDER (5450 Å)	COLOR AND COMMENTS	FILM THICKNESS (MICROMETERS)	ORDER (5450 Å)	COLOR AND COMMENTS
0.050		tan	0.60		carnation pink
0.075 750 Å		brown	0.63		violet red
			0.68		"bluish"**
0.100		dark violet to red violet			
0.125		royal blue	0.72	IV	blue green to green (quite broad
0.150		light blue to metallic blue	0.77		"yellowish"
0.175	I	metallic to very light yellow green			
			0.80		orange (rather broad for orange)
			0.82		salmon
0.200		light gold or yellow—slightly metallic	0.85		dull, light red violet
			0.86		violet
0.225		gold with slight yellow orange	0.87		blue violet
0.250		orange to melon	0.89		blue
0.275		red violet			
			0.92		blue green
0.300		blue to violet blue	0.95	V	dull yellow green
0.310		blue	0.97		yellow to "yellowish"
0.325		blue to blue green	0.99		orange
0.345		light green			
0.350		green to yellow green	1.00		carnation pink
			1.02		violet red

Thickness	Order	Color	Thickness	Order	Color
0.365		yellow green	1.05		red violet
0.375	II	green yellow	1.06		violet
0.390		yellow	1.07		blue violet
0.412		light orange	1.10		green
0.426		carnation pink	1.11		yellow green
0.443		violet red	1.12	VI	green
0.465		red violet	1.18		violet
0.476		violet	1.19		red violet
0.480		blue violet	1.21		violet red
0.493		blue	1.24		carnation, pink to salmon
0.502		blue green	1.25		orange
0.520		green (broad)	1.28		"yellowish"
0.540	III	yellow green	1.32	VII	sky blue to green blue
0.560		green yellow	1.40		orange
0.574		yellow to "yellowish"*	1.45		violet
0.585		light orange or yellow to pink borderline	1.46		blue violet
			1.50	VIII	blue
			1.54		dull yellow green

*Not yellow, but is in the position where yellow is to be expected; at times it appears to be light creamy grey or metallic.
**Not blue but borderline between violet and blue green; it appears more like a mixture between violet red and blue green and overall looks greyish.
NOTE: Above chart may also be used for Vapox, Silox, and other deposited oxide films. For silicon nitride films, multiply film thickness by 0.75.
SOURCE: *IBM J. Res. Dev.*, 8, 43 (1964).

3.4 | Oxide Evaluation

The two important characteristics of a layer of SiO_2 are its thickness and its dielectric quality. The thickness can be accurately predicted from the oxidation sequence, but it is often necessary to verify the results. The thin, uniform layers of SiO_2 resulting from oxidation appear to have colors when observed perpendicularly in white light. A chart of oxide colors and associated oxide thicknesses is given in Table 3–1. As the oxide thickness increases from no oxide on a bare wafer to a very thick layer, the colors observed with these thicknesses repeat themselves. To determine the thickness of a layer of oxide from its color, you must also know all the colors preceding it on the color chart. This requirement is easily met by etching a taper in the layer of oxide by immersing the layer in hydrofluoric acid and slowly withdrawing it (Figure 3–6). The thickness can also be determined electrically by measuring the capacitance it produces between two conductive plates of known area, or by using optical interference or physical techniques. An instrument often used to determine the thickness of a layer is a *surface profilometer*. A stylus with an electronically amplified output is drawn over the surface to be measured, and an ink trace of the resulting profile shows the height of any steps in the surface. The instrument is used to determine the thickness of a layer of SiO_2 once a step has been obtained by etching.

Very accurate determinations of oxide thickness are made using an optical instrument called an *ellipsometer*. The ellipsometer uses a beam of light that travels through the SiO_2 layer to determine its thickness. However, the degree of accuracy that is attainable using this instrument is necessary in only a few instances, such as MOS gate oxide thicknesses.

The dielectric quality of a layer of SiO_2 is usually determined by two parameters:

1. The breakdown strength of a layer of SiO_2.
2. The amount of contamination present in the SiO_2 layer that drifts when a voltage is applied across the layer.

The first parameter is measured by placing a voltage between two conducting plates across a known thickness of SiO_2 and increasing the voltage until the

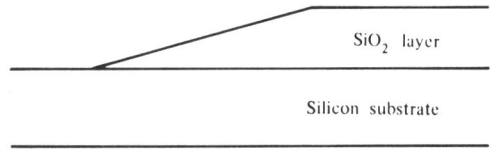

Figure 3–6: Tapered oxide thickness on a silicon wafer.

current between the two plates increases significantly. The breakdown value is measured for a predetermined number of areas on a wafer, with the quality of the oxide determined by the breakdown voltage distribution. For silicon dioxide, 600 V/micrometers is considered an acceptable dielectric strength.

The presence of mobile contamination (generally sodium) is measured using a technique called Capacitance–Voltage, or C–V. This analysis technique uses a thin layer of SiO_2 with conducting plates on either side. The capacitance of the structure is measured after it has been biased first negative at one terminal with respect to the other, and then positive at one terminal with respect to the other at an elevated temperature. The voltage shift between the C–V curves at these extremes is related to the amount of mobile contamination. A typical C–V curve is shown in Figure 3–7.

3.5 Oxidation Technology for Low Contamination Levels

Recent studies have indicated that the measured dielectric quality of a layer of SiO_2 is increased and the apparent amount of mobile contamination is decreased if a chlorine-carrying compound such as HCl or TCE is injected into the oxidation tube during the growth of the oxide layer. However, the amount of the chlorine-carrying species must be controlled within certain bounds to obtain the benefits. It has been hypothesized that the chlorine ions accumulate near the Si–SiO_2 interface, where they combine with any mobile contamination they encounter, rendering the impurity immobile. Similarly, during an oxidation cycle they may tie up any impure material they encounter.

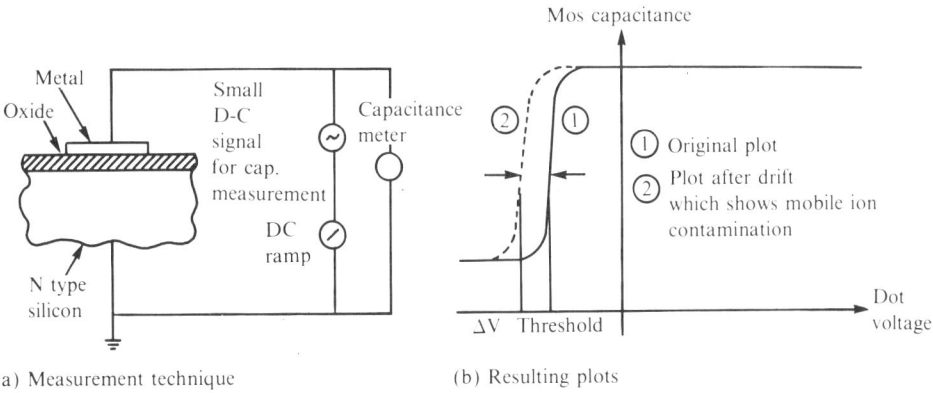

Figure 3–7: A typical capacitance-voltage (C-V) set-up and resulting curve showing the presence of mobile ionic contamination.

3.6 | Oxidation Reaction

An understanding of the physical process that takes place during oxidation is obtained by considering the two extreme cases of a reaction, namely, (1) the transport-limited case, and (2) the reaction-rate-limited case. (See Figure 3–8.) Once any amount of oxide has grown on the surface of a silicon wafer, the oxidizing species must move through this oxide layer to reach the silicon. The oxide growth can be limited by:

1. The availability of the oxidizing species at the Si–SiO$_2$ interface.
2. The ability of the reaction between the oxidizing species and the silicon to take place.

The presence of a very thin layer of SiO$_2$ does not interfere with the diffusion of oxidizing atoms to the Si–SiO$_2$ interface. These atoms diffuse to the interface until an excess of them is present. The SiO$_2$ growth rate is limited by the speed with which the silicon can react with the oxidizing atoms. This case, called the reaction-rate-limited case, is shown in Figure 3–8a. When the oxide layer is sufficiently thick, the oxidizing species cannot diffuse through this layer rapidly enough to keep the reaction going at peak speed. This case, the "transport-limited" or "diffusion-limited" case, is shown in Figure 3–8b.

The different growth rates in the reaction-rate-limited and transport-limited cases are illustrated in Figures 3–9 and 3–10. For short growth times at low temperatures, the slope of the oxide-thickness-versus-time curve is different from that for high temperatures and long growth times. The steeper slope corresponds to the reaction-rate-limited case, the less steep slope to the transport-limited case.

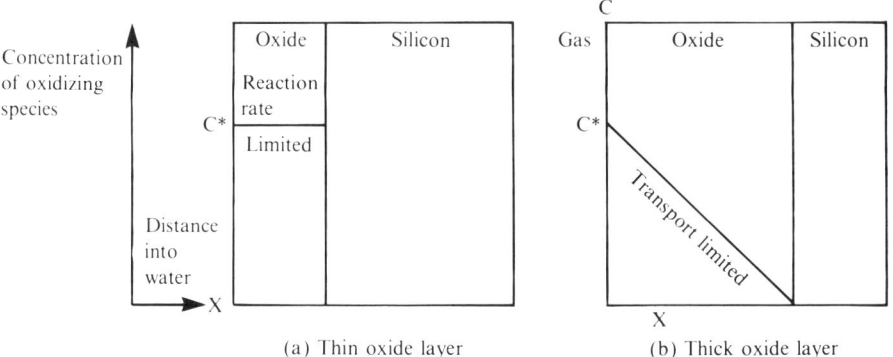

Figure 3–8: Distribution of the oxidizing species in the oxide layer for the two limiting cases of oxidation.

3.7 Oxide Thickness Determination

Though both oxygen and water vapor are used to oxidize silicon wafers, oxidations using these two species are not interchangeable. Oxidation using water vapor proceeds at a significantly faster rate than oxidation using oxygen, for the same temperature and time. The different oxidation rates for the two species give rise to a different set of applications for each type of oxidation. The oxide thickness resulting from a single oxidation step starting with a bare sili-

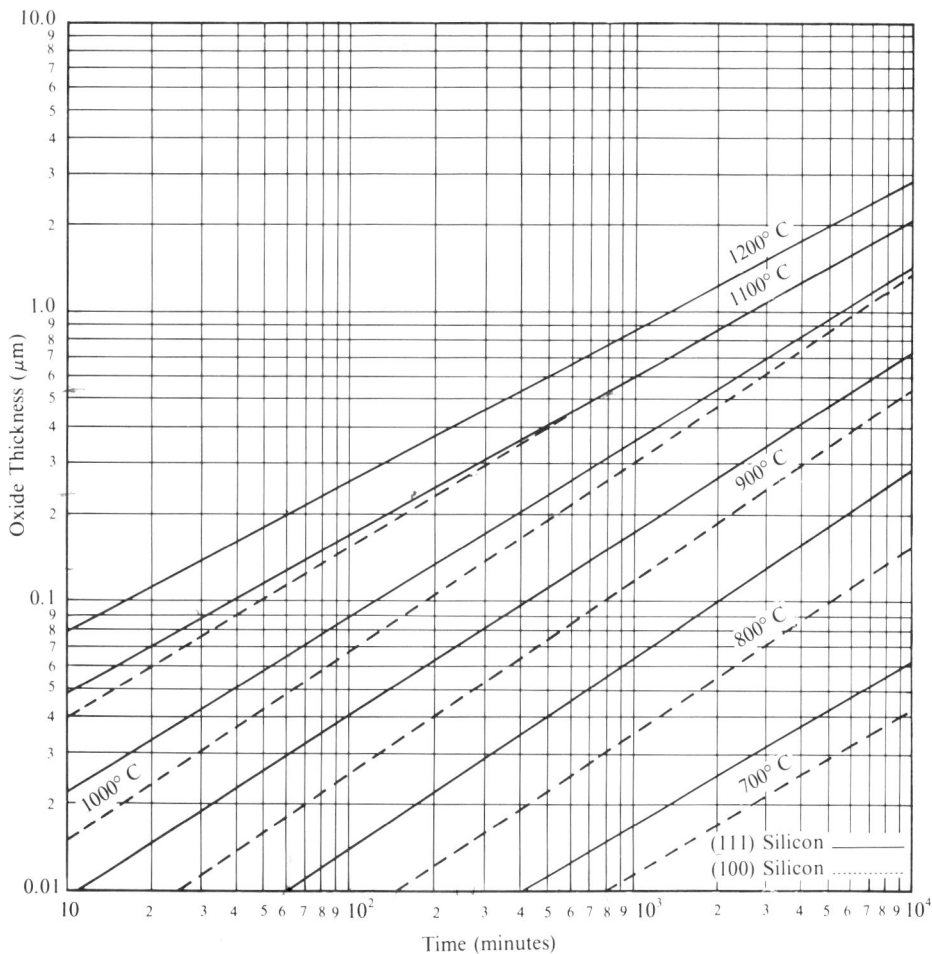

Figure 3–9: Oxide thickness versus oxidation time for silicon in dry O_2. *(Ref. 1.)*

con wafer is determined using Figures 3–9 and 3–10. Some specific examples of oxide thickness determination are given below:

EXAMPLE 1

Determine the SiO_2 thickness following a 70-minute, 900°C steam oxidation cycle on <111> silicon.

SOLUTION

Using Figure 3–10, find the line labeled 900°C and determine its intersection with the line denoting 70 minutes. From this point of intersection, proceed directly across to the left-hand edge of the graph. The thickness indicated is 2300 Angstrom units (Å) or .23 μm.

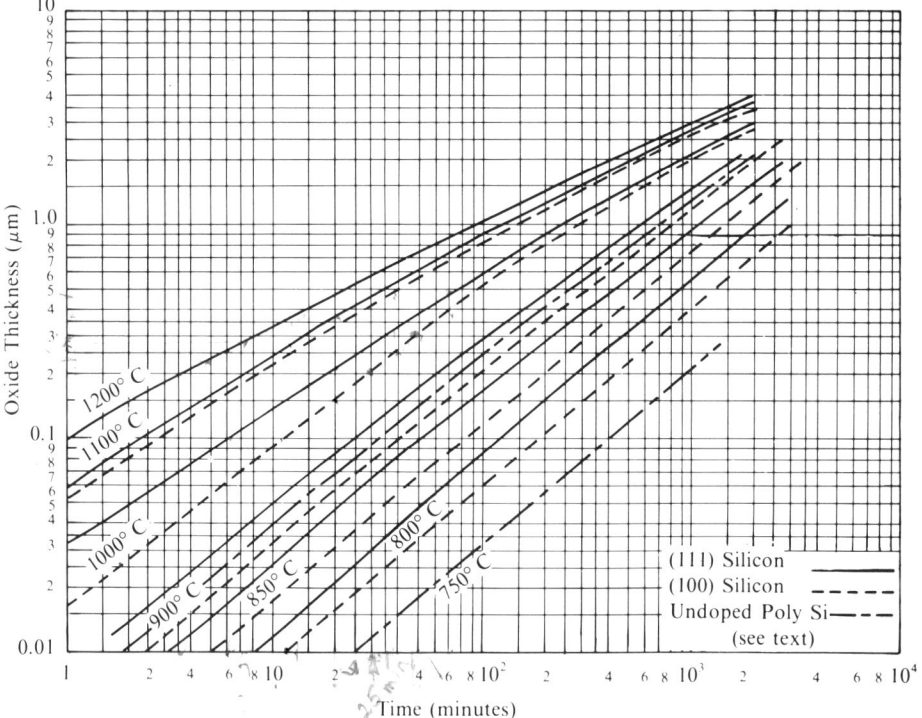

Figure 3–10: Oxide thickness versus oxidation time for silicon in pyrogenic steam (640 Torr). *(Ref. 1.)*

3.7 Oxide Thickness Determination

EXAMPLE 2

Determine the SiO_2 thickness following a 90-minute, 1000°C dry O_2 oxidation cycle on <111> silicon.

SOLUTION

Using Figure 3–9, find the line labeled 1000°C and determine its intersection with the vertical line representing 90 minutes. From this point of intersection, proceed directly across to the left-hand edge of the graph. The intersection along this axis is 830Å or .083 μm.

In typical device processing, a silicon wafer sees a sequence of oxidation cycles in which both the temperature and the oxidizing species vary. The thickness of an oxide layer following any series of oxidation steps is still determinable as long as these rules are followed:

1. Always begin with the present thickness of SiO_2. Determine how long it takes to grow this thickness using the next-growth conditions. (If the wafer is bare, the time is zero.)

2. To the amount of time determined in step 1, add the additional oxidation time in the present cycle.

3. Find the oxide thickness resulting from the time in step 2. If the cycle was the last oxidation cycle, then you have the total thickness. If not, use this thickness as the starting point and go through steps 1, 2, and 3 again.

As an example of this technique, consider the following sequence of oxidations on <111> silicon:

1. Dry O_2 @ 1200°C for 60 minutes.
2. Steam @ 900°C for 45 minutes.
3. Steam @ 1200°C for 15 minutes.

The oxide thickness following the first cycle is determined directly from Figure 3–9. We see from this figure that the thickness of the first cycle is 2000 Å. Now, using this as the starting thickness, determine how long it takes to grow the 2000-Å layer using the growth condition of the second oxidation step. The next oxidation takes place at 900°C in steam. Find the point at which the 2000-Å line crosses the 900°C growth line on Figure 3–10. From the figure, the time to grow 2000 Å of SiO_2 at 900°C is 60 minutes. As far as the silicon is concerned, the 2000 Å of SiO_2 present on the surface of the wafer could have taken 60 minutes to grow at 900°C. We next add another 45 minutes at 900°C to the growth time, resulting in the equivalent of 105 minutes at 900°C.

Again, from the figure, 900°C for 105 minutes at the stated growth conditions results in 3000 Å of SiO_2.

Now, with 3000 Å as a starting point, follow the procedure once more. Determine the point at which the 3000-Å line intersects the 1200°C growth curve on Figure 3–10. The figure shows that it would have taken 8 minutes to grow 3000 Å at this growth condition. This 8 minutes is thus used as the starting time, and adding the additional 15 minutes at 1200°C, we end up with 23 minutes at 1200°C in steam. Looking to the 1200°C growth curve, we see an intersection at the 5000-Å line, so the final oxide thickness is 5000 Å.

3.8 Redistribution of Dopant Atoms During Thermal Oxidation

During the course of thermal oxidation, the interface between the silicon layer and the silicon dioxide layer moves through doped regions of the silicon. Dopants present at this moving interface redistribute themselves depending on

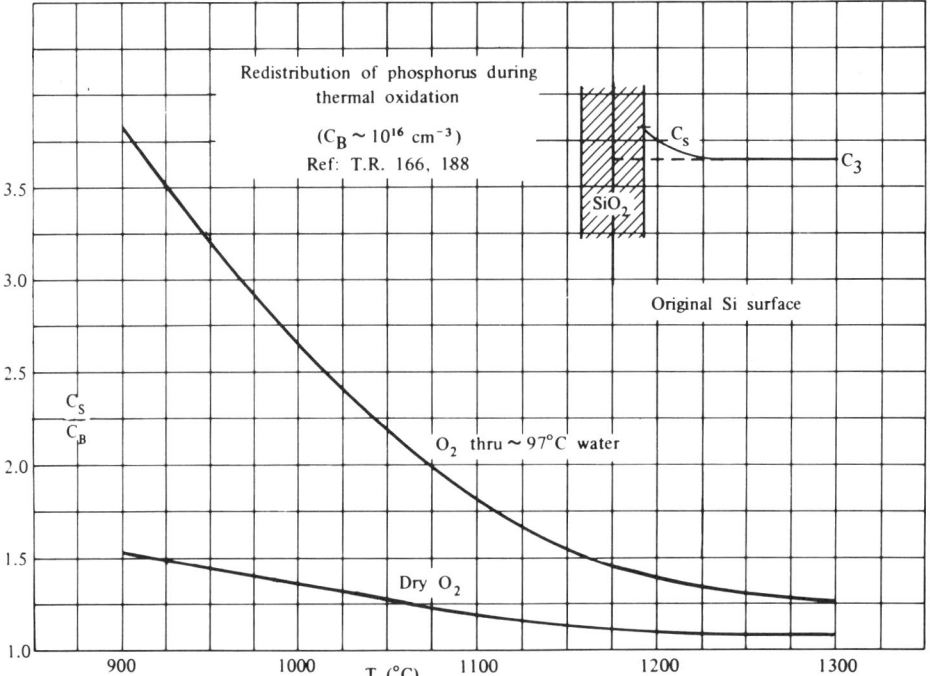

Figure 3–11: Redistribution of phosphorus during thermal oxidation. (Ref. 2.)

their relative solubility in silicon and silicon dioxide. Phosphorus, arsenic, and antimony have a greater solubility in silicon than in silicon dioxide, so these impurities tend to pile up in front of an advancing Si–SiO$_2$ interface. Boron, on the other hand, has a greater solubility in silicon dioxide, so boron is depleted from the silicon in front of an advancing Si–SiO$_2$ interface and accumulates in the newly grown silicon dioxide layer. The redistributions of phosphorus and boron during thermal oxidation are shown in Figures 3–11 and 3–12, respectively.

3.9 | High-Pressure Oxidation

High temperatures and long times are required to grow thick oxide layers using atmospheric growth techniques. Some fabrication sequences, particularly those requiring shallow junctions, are not compatible with these oxidation cycles. High-pressure oxidation is one technique for growing an oxide layer without needing a long, high-temperature cycle.

A high-pressure oxidation cycle uses water vapor for the oxidizing species and occurs in a specially designed furnace. The pressure range used varies but

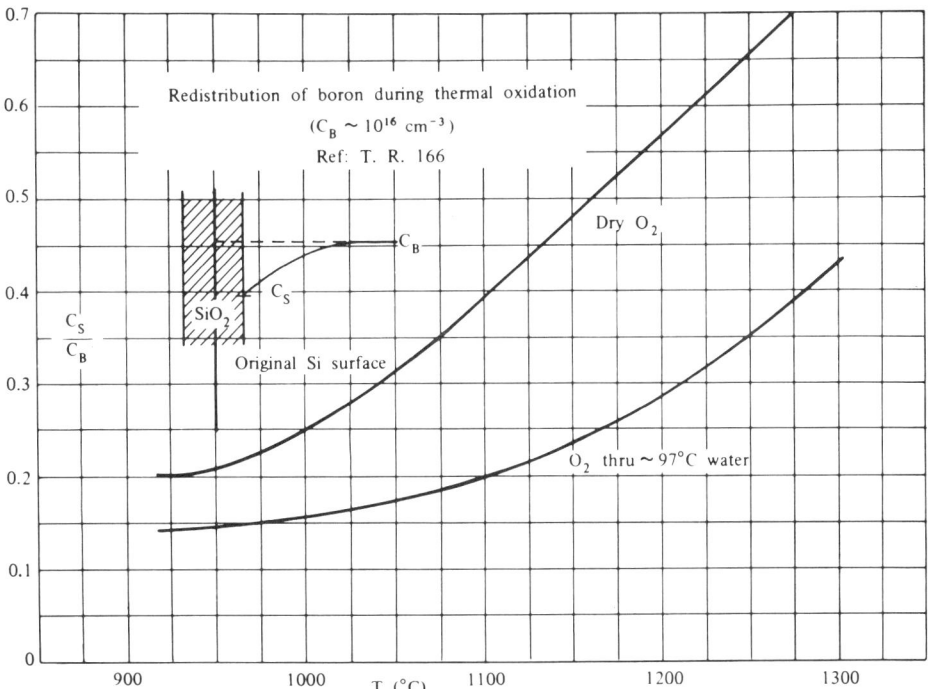

Figure 3–12: Redistribution of boron during thermal oxidation. *(Ref. 2.)*

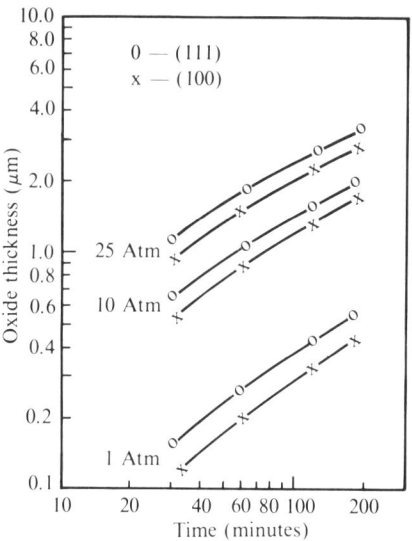

Figure 3–13: High-pressure steam oxidations; 920°C. *(Ref. 3.)*

is typically less than 50 atmospheres. The presence of the water vapor at high pressure at the surface of the wafer increases the amount of water that dissolves in the SiO_2. The higher surface concentration results in a greater amount of water moving through the SiO_2 layer to the Si–SiO_2 interface, where growth occurs. The thickness of a layer of SiO_2 resulting from various high-pressure growth cycles is shown in Figure 3–13.

3.10 | Anodic Oxidation

Anodic oxidation is a technique for growing a relatively thin layer of SiO_2 (up to about 600 Å in thickness) on a wafer at low temperatures. In this technique, a positive voltage is applied to a silicon wafer in an electrolytic solution, mak-

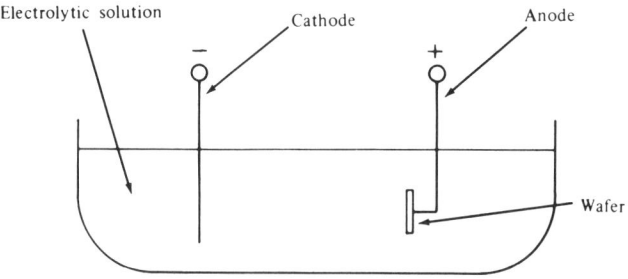

Figure 3–14: Apparatus for the anodic oxidation of silicon.

ing the wafer the anode. Then a cathode is placed in the solution, and a voltage is applied from anode to cathode (Figure 3–14). This voltage determines the final oxide thickness on the wafer. The oxide layer formed using this technique is of poor quality from an electrical standpoint, but silicon layers of reproducible thickness are easily removed because the silicon diffuses through the oxide layer to the SiO_2–electrolyte interface, leaving the profile of any previously diffused dopant unchanged. A combination of anodic oxidation, SiO_2 removal, and four-point probing is sometimes used to determine the profile of a diffusion.

Review Exercises

1. Determine the thickness of a layer of SiO_2 grown on a bare <111> silicon wafer if the oxidation cycle is:
   a. Dry O_2 @ 1200°C for 60 minutes.
   b. Steam @ 1000°C for 12 minutes.
2. If the time a wafer is oxidized is doubled, does the oxide thickness double? If not, why not?
3. Determine the thickness of a layer of SiO_2 grown on a bare <111> silicon wafer in the following three sequential steps:
   a. Dry O_2 @ 1100°C for 4.5 hours.
   b. Steam @ 1200°C for 6 minutes.
   c. Steam @ 1100°C for 12 minutes.
4. Is silicon or the oxidizing species (either O_2 or H_2O) the mobile species in anodic oxidation?
5. A <100> silicon wafer is oxidized for 24 minutes in steam at 1100°C. How much time is needed to grow an additional 500 Å of oxide in dry O_2 at 1000°C, and what is now the total oxide thickness?
6. Explain why steam oxidation of silicon proceeds at a faster rate than dry oxidation.
7. Explain the difference between transport-limited and reaction-rate-limited oxidation.
8. When boron-doped silicon is oxidized, is the tendency for the boron in the silicon to pile up or be depleted at the silicon–oxide interface? Explain your reasoning.

REFERENCES (CHAPTER 3)

1. O. D. Trapp et al., *Semiconductor Technology Handbook*, Technology Associates, 1985, pp. 4.1–4.2. (This book is available from Technology Associates, 51 Hillbrook Dr., Portola Valley, Ca. 94025.)

2. A. S. Grove, *Physics and Technology of Semiconductor Devices*, John Wiley and Sons, 1967, pp. 72–73.

3. L. E. Katz, et al., *Solid State Technology*, December, 1981, p. 87.

4 | Photolithography

4.0 | Introduction

The photolithographic process in the semiconductor fabrication sequence is the step during which the geometric pattern that produces the desired electrical behavior is transferred to the surface of the wafer. The word "photolithography" may be loosely defined as "printing with light," which is an accurate description of the heart of this process step. The manufacture of semiconductor devices and integrated circuits consists of multiple passes through photolithography (or "masking" as it is often called), with process steps such as impurity introduction, oxidation, or metallization following each masking step. The masking step defines the region where the subsequent process step will have its effect. The sum of all process steps produces devices and circuits with specific electrical behavior. This chapter discusses the photolithographic process and its importance in the manufacture of semiconductors.

4.1 | Process Overview

The complexity of the photomasking operation is best introduced by discussing the key process steps prior to beginning a more complete discussion. These steps are shown sequentially in Figure 4–1. In Figure 4–1a, a photomask with a geometric pattern as its lower surface is in contact with an oxidized silicon wafer that has been coated with photoresist. *Photoresist* is a *photo*sensitive, etch-*resistant* material used to transfer the image from the mask to the layer of silicon dioxide on the surface of the wafer. In the figure, the photoresist is "exposed" using a source of intense ultraviolet light such as a mercury arc

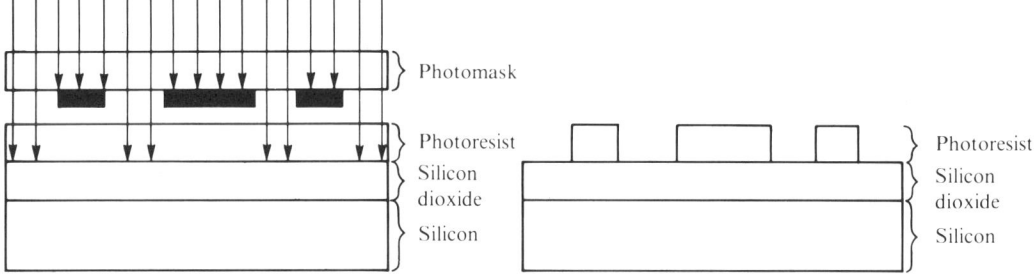

(a) A photomask is positioned above an oxidized wafer that has been coated with positive photoresist and it is exposed using an intense light source.

(b) The mask is removed and the photoresist coated wafer is developed. With positive resist, the developer removes the resist where the light has "exposed" the resist, leaving a positive image in the resist.

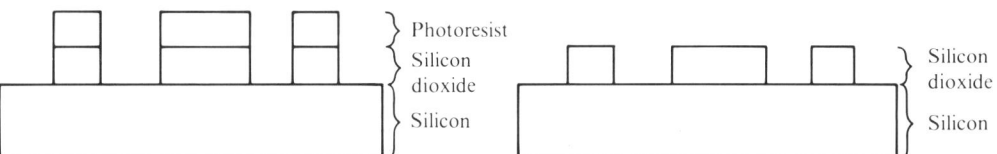

(c) The photoresist is used to mask region of the underlying silicon dioxide layer during an etch step that removes unprotected oxide.

(d) The remaining photoresist is removed, leaving the image of the mask etched in the silicon dioxide layer.

Figure 4–1: The transfer of a device geometry from a photomask to the surface of a wafer using positive photoresist. (a) A photomask is positioned above an oxidized wafer that has been coated with positive photoresist and it is exposed using an intense light source. (b) The mask is removed and the photoresist-coated wafer is developed. With positive resist, the developer removes the resist where the light has "exposed" it, leaving a positive image in the resist. (c) The photoresist is used to mask regions of the underlying silicon dioxide layer during an etch step that removes unprotected oxide. (d) The remaining photoresist is removed, leaving the image of the mask etched in the silicon dioxide layer.

lamp. The photoresist shown is a positive photoresist, and the intense light alters its chemical bonding to make it more soluble where it has been exposed. In Figure 4–1b, the wafer is removed from the alignment station and "developed." The exposed regions of the positive photoresist layer are dissolved in a developing solution, leaving a pattern that is a duplicate or "positive" of the mask on the surface of the silicon dioxide layer. This image-transfer step is the first of the two image-transfer processes (mask to photoresist, and photoresist to silicon dioxide) that take place in the photomasking step.

In the steps illustrated in Figures 4–1a and 4–1b, the photosensitive property of photoresist has been used. Its etch-resistant property is used in the etching process shown in Figure 4–1c. In this process, an etchant (a liquid or a gas) is chosen to remove the silicon dioxide in regions that are not protected by the remaining photoresist. If properly selected, the etchant will remove the layer of silicon dioxide but will not etch the underlying silicon or the layer of photoresist, as shown in the figure. The result of the photolithographic process is shown in Figure 4-1d where, after the layer of resist has been removed, only the patterned layer of silicon dioxide is left. The second of the image-transfer steps—from the photoresist layer to the layer of silicon dioxide—has thus been completed.

4.2 Photoresist: The Key to Image Transfer

The photolithographic process is the transfer of an image from the mask to a wafer through the use of a photosensitive material often called photoresist. Photoresist is a chemical substance containing a light-sensitive material in suspension in a solvent. The light-sensitive material is selected so that it responds to the intense blue-violet light put out by a mercury arc lamp, but does not respond to the red or yellow light commonly in use in darkrooms or photoresist areas. Photoresist comes in two distinct types:

1. *Positive resist.* The light from the exposure step increases the solubility of the resist in the developing solution. This type of resist may be thought of as "photo- or light-softened" resist.

2. *Negative resist.* The light from the exposure step causes a process called polymerization to occur in the resist. (The intense energy used for exposure causes chemical cross-leaking between strands of resist, producing a stable sheet of resist in these areas.) This type of resist may be thought of as "photo- or light-hardened" resist.

Either type of resist may be used to etch the films and materials used in semiconductor manufacturing. Negative photoresist was more popular when the linewidths were larger. As the linewidths and spacings became smaller (5 micrometers is often thought of as the cross-over dimension), the ability of positive resist to transfer smaller patterns made it more popular. Both positive and negative resists are complex organic molecules containing carbon, oxygen, nitrogen, and hydrogen.

A photoresist is characterized by four parameters that affect its performance:

1. *Adhesion:* A measure of the lateral etch at the edge of a post-baked resist image.

2. *Etch resistance.* An oxidized wafer fully coated with photoresist is subjected to an etch several times longer than normal. The observer looks for any breakdown in the photoresist.

3. *Resolution.* The minimum width and spacing that can be successfully transferred to the resist layer.
4. *Photosensitivity.* The absolute response to different light intensities.

These tests are performed by a manufacturer to maintain a constant production quality. The tests may also be performed by the user on all lots of photoresist, to verify that the high-quality photoresist necessary for today's manufacturing is always being used.

The amount of solvent in a basic photoresist determines its thickness or viscosity. The more viscous a resist, the less easily it flows. (Honey is a more viscous material than water because it does not spread as fast as water does from a drop on a surface.) The viscosity of a photoresist is measured in units of either centipoise or centistoke. These units are closely related but are not the same. Most photoresists are used in the 14–60 centipoise range which means that their flow characteristics are similar to those of syrup.

4.3 The Photolithographic Process Sequence

The photolithographic process consists of a number of steps performed sequentially, regardless of the particular photoresist being used or the layer to which it is being applied. These steps are shown in the following photoresist flowchart and subsequently described.

Basic Photoresist Flowchart

STEP	OPERATION
Prepare substrate.	Oxidize, perform CVD, metallize.
Prepare surface.	Clean, blow with N_2, dehydrate, prime, bake.
Apply resist.	Spin, spray, dip.
Soft-bake.	Cure at low temperature to dry the resist.
Align and expose.	Align and selectively expose the resist.
Develop resist.	Dissolve the resist in selected regions.
Inspect development (develop check).	Verify that accurate image has been transferred to the photoresist.
Hard-bake.	Cure at higher temperature to completely dry the resist and ensure adequate adhesion.
Etch.	Use plasma or a wet etchant to etch away regions not protected by resist.
Strip resist.	Remove resist organically, or by asher, plasma, or acid.
Inspect visually (final inspection).	Verify accurate image transfer to the layer before sending to the next step.

4.3 The Photolithographic Process Sequence

PREPARE SUBSTRATE

Both the layer to be etched and its surface conditions are determined by this step, which also dictates the exact processes that will be used in the photoresist flow.

PREPARE SURFACE

In many cases, such as wafers that have just been removed from a diffusion or oxidation furnace or from a metal evaporator, no surface preparation is needed. However, some surfaces, such as silicon nitride or polycrystalline silicon, may require preparation. Oxidation of both of these materials is a common technique and is performed as described in the section on oxidation. Another technique that is sometimes used is called *priming*. The use of a priming solution increases the adhesion of the photoresist to the surface. Primer may be applied by immersing the substrates in the priming solution, by spraying the solution on, or by passing gas laden with priming vapors over the surface of the wafer. Application of some primers requires baking the substrates before subsequent coating with photoresist. A dehydration bake at temperatures above 100°C (the boiling point of water) may be used if the wafers have been in an area with high relative humidity.

APPLY RESIST

Photoresist may be applied using a variety of techniques, including dipping, spraying, brushing, or rollercoating. In the fabrication of semiconductor devices, the most satisfactory technique is the use of a "spinner." A cross section of a spinner with a wafer on it is shown in Figure 4–2. The apparatus consists of a wafer chuck on a hollow shaft, both of which rotate. A vacuum holds the wafer on the chuck while it is in motion. To apply the resist, a precise amount of it is dispensed onto the center of the wafer, and the wafer begins to spin. As the wafer spins, the photoresist on it moves outward, uniformly coating the wafer. Excess photoresist is spun off the edge of the wafer. The spin speed and the viscosity of the resist determine the thickness of the photoresist following application. A typical spin-speed-versus-time diagram is shown in Figure 4–3.

The five distinct periods during application of the resist are:

1. *Initial acceleration*. The wafer begins spinning and is accelerated to a constant, but relatively low, spin speed. The resist is usually applied before the wafer begins spinning.
2. *Resist spread*. The low spin speed guarantees that the photoresist is spread uniformly across the wafer. Any excess resist is thrown off the wafer.

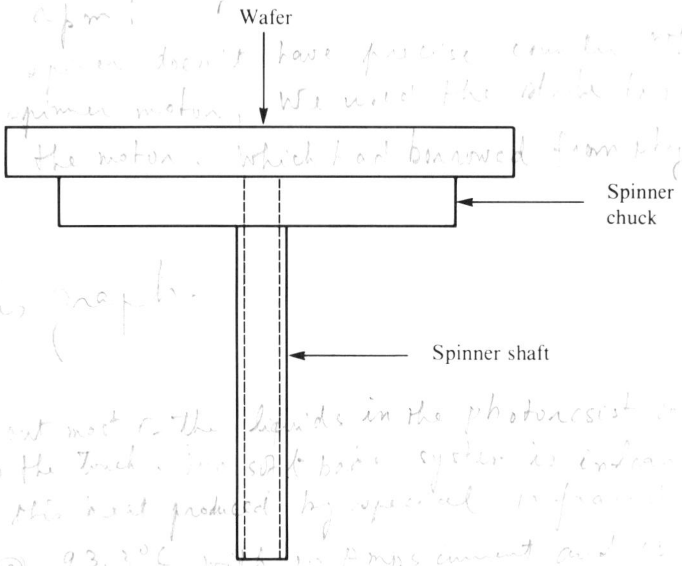

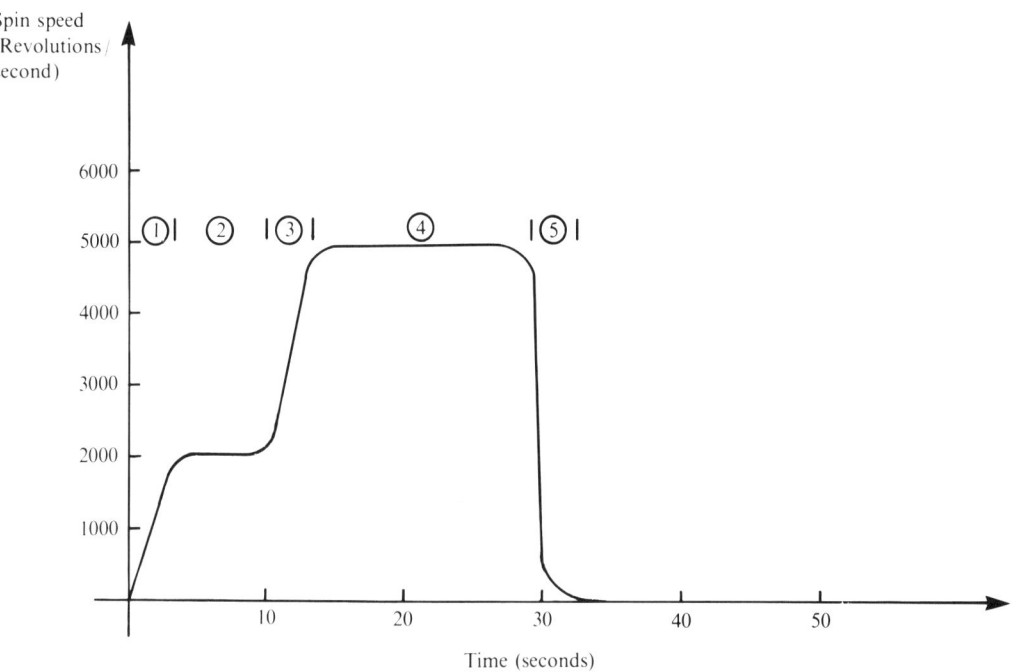

Figure 4–2: Cross section of a spinner with a wafer on it.

Figure 4–3: Spin speed versus time during a typical resist application cycle.

4.3 The Photolithographic Process Sequence

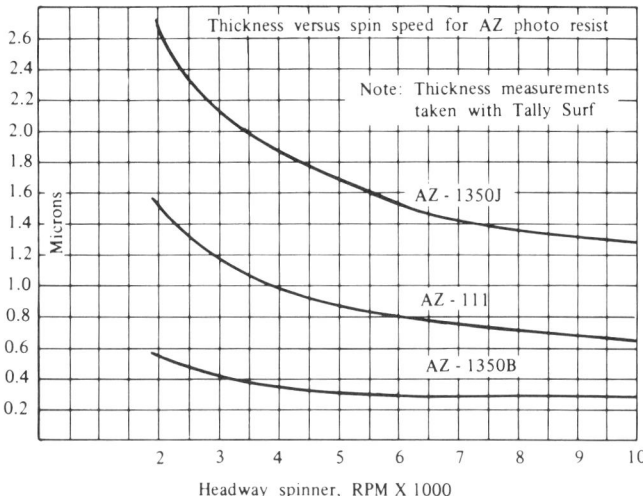

Figure 4-4: Photoresist thickness vs. spin speed for different viscosity photoresists.

3. *Final acceleration.* The spin speed is rapidly increased to the final spin speed.

4. *Final spin.* The speed of the wafer during this operation determines the thickness of the resist. A higher spin speed results in a thinner resist layer, as shown in Figure 4-4.

5. *Deceleration.* The velocity of the wafer is reduced to zero so that the wafer may be removed and the cycle started again with another wafer.

A photoresist has specified minimum and maximum spin speeds for obtaining uniform layers. If too low a spin speed is used, an excessive edge bead forms because the resist is not thrown from the wafer. Too high a spin speed produces a nonuniform layer because of uneven spreading and uneven evaporation of the solvent in the resist.

SOFT-BAKE

Following resist application, excess solvent is baked out of the resist during the soft-bake step. Methods of baking the resist that are in common use include:

1. *Convection heating.* A current of hot filtered air or dry nitrogen circulates through an oven set at a constant temperature, evaporating the solvent.

2. *Infrared (IR) heating.* The heat produced by special infrared bulbs heats the wafers, evaporating the excess solvent.

3. *Conductive heating.* The wafer sits directly on a heated plate or wafer chuck, heating the wafer and evaporating the solvent.

(Other techniques of soft baking, including microwave and inductive heating, have been investigated, but the results have not proven as satisfactory as those obtained with the preceding three methods.)

Temperature and time are the two major variables that are controlled. Baking at too low a temperature requires excessive time, while baking at too high a temperature results in the surface being sealed while solvent is still present in subsurface levels. This condition leads to a wrinkled appearance in the resist surface. The specific time and temperature are set by the photoresist being used, the method of baking, and the layer the resist is coating. Typical times and temperatures are 80°C to 100°C for 10 to 20 minutes.

ALIGN AND EXPOSE

After the wafers have been cooled, they are ready to be aligned and exposed. An aligner, a precise piece of opto-mechanical or optical-mechanical equipment, is used for this purpose. The mask and the wafer are positioned in a precise manner with respect to each other. If the wafer already has a pattern on its surface (that is, if it has already gone through first mask), the mask is aligned to the pattern that exists. If the mask is the first, the orientation with respect to wafer flats is important. The alignment of successive layers is made possible by magnifying optics and precise controls for positioning the wafer with respect to the mask.

The semiconductor industry has undergone an evolution in mask alignment technology in the last 20 years. Driven by the need for smaller geometries on larger wafers, this evolution has given rise to the following four types of aligners, shown in Figure 4–5:

1. *Contact printer.* The contact printer places the mask in direct contact with the wafer after the mask is aligned. The entire wafer is then exposed at one time. As seen in Table 4–1, contact printing is the least expensive technique, but it does not produce the fine pattern sizes needed as technology progresses. In addition, the mask quality deteriorates with use, resulting in more defects as the number of contacts between the mask and wafers is increased. This alignment method is used most frequently for small die-size devices and circuits where the defect density does not affect the yield too significantly.

2. *Proximity printer.* The proximity printer is similar to the contact printer, except that it doesn't clamp the mask to the wafer. This significant difference results in much longer mask life. However, the minimum feature size is also relatively large, as seen in Table 4–1. Proximity alignment is

4.3 The Photolithographic Process Sequence

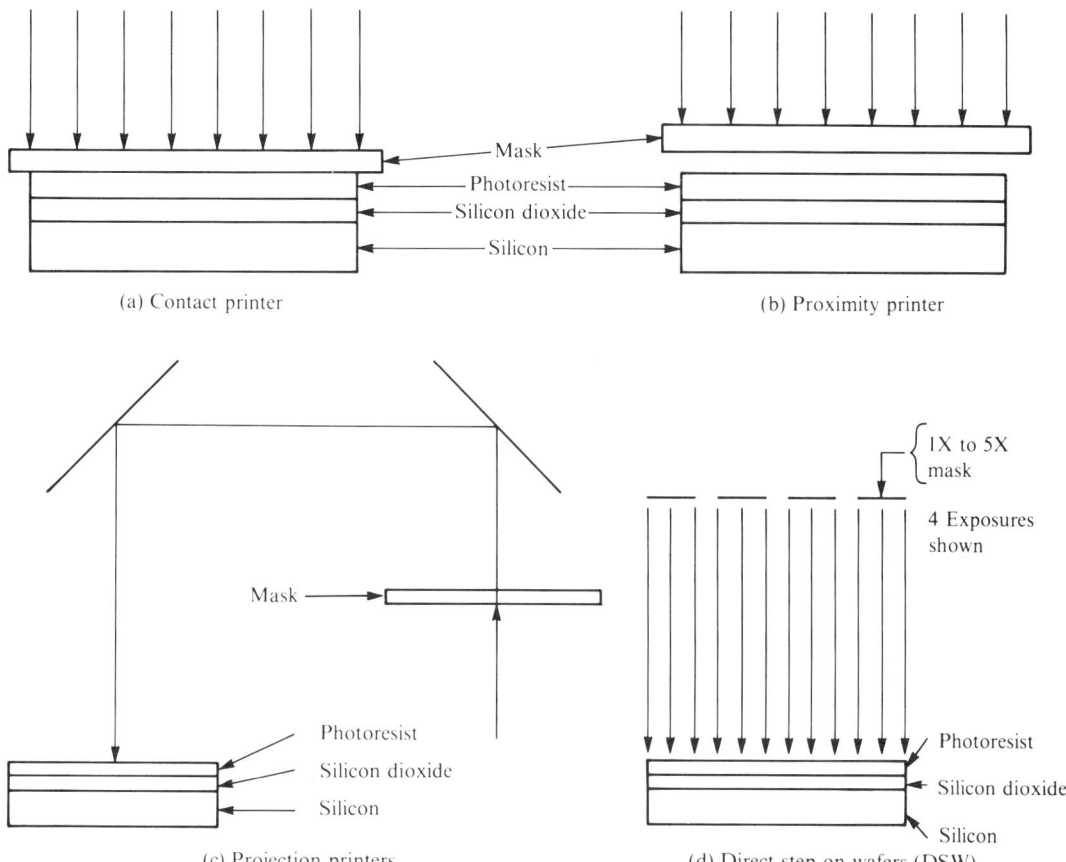

Figure 4–5: The types of mask aligners used in semiconductor manufacture. (a) Contact printer; (b) Proximity printer; (c) Projection printers; (d) Direct step-on wafers (DSWs).

used where small feature sizes are not required, such as on older discrete devices and ICs, and on pad mask.

3. *Projection printer.* In contrast to contact and proximity printers, which place the mask close to the wafer being exposed, the projection printer leaves the wafer and the mask separated by some distance. In addition, the optics of this type of aligner do not allow the entire wafer to be exposed at one time. Instead, the pattern on the mask is swept across the wafer, exposing an arc-shaped region.

4. *Direct step-on wafer (DSW).* Instead of creating a mask that is the same size as the wafer, DSW uses a mask that contains many fewer circuit or device patterns than are on the wafer. These patterns are typically $1\times$, $5\times$, or $10\times$ magnifications of the pattern that is obtained on the wafer.

Table 4–1: Comparison of Mask Alignment/Expose Equipment

ALIGNMENT TECHNIQUE	MASK-TO-PATTERN RATIO	MINIMUM FEATURE SIZE (μm)	WAFERS PER HOUR	COST RANGE (THOUSANDS OF $)
Content	1:1	5	60	40–80
Proximity	1:1	4	60	80–150
Projection	1:1	2.5	40	200–400
DSW	1:1 to 10:1	1.25	20	300–600

The mask is aligned to part of the wafer and then exposed before moving or "stepping" on to repeat the process.

The intense ultraviolet radiation of a mercury arc lamp is often the source of the light for the exposure step. A Xenon flash tube, another source of intense light, is occasionally used. However, the energy from either lamp is sufficient to cause many negative resists to react with oxygen, so the wafer surface is blanketed with nitrogen when negative resist is used to prevent this reaction.

DEVELOP RESIST

Following alignment and exposure, those photoresist regions with the highest solubility are dissolved "developing" the pattern. (For negative resist, the unexposed regions are dissolved, for positive resist, the exposed regions.) Development may be performed by immersing the wafer in the developer or spraying the developer on, either with or without having first atomized it. Atomizing the developer uses a minimum amount of developer and is favored in many applications. Development should leave a sharp edge where the photoresist stops. A rinse is usually applied following development to remove any residual material. When development is automated, the developer and the rinse are dispersed, so that there is a slight amount of overlap in the cycles, as

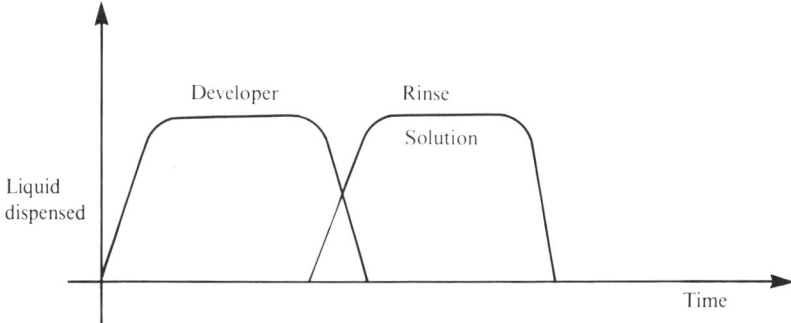

Figure 4–6: A typical development cycle, showing the overlap of the developer and the rinse solution.

shown in Figure 4–6. This overlap prevents any residue from drying on the surface of the wafer.

The chemicals used for developing and rinsing are chosen for the specific resist in use.

DEVELOP INSPECT

Following development, the wafer is examined to determine the accuracy of the first image-transfer process. Manual inspection is performed by an operator scanning the surface of the wafer looking for any problems with the remaining resist pattern. Automated equipment is also available to perform this function. Wafers that do not pass this strict inspection are removed from the run and are reworked. (Sending wafers on with known resist problems produces wafers with the same defects.)

HARD-BAKE

The purpose of hard baking is to evaporate any solvents left in the resist layer following development and to increase adhesion at the edges of the photoresist pattern. The task is accomplished using any of the soft-baking methods, but set at a higher temperature. (Some positive resists may not require hard baking; the details of processing determine the matter.) A typical hard-bake cycle is 10 to 20 minutes at 110–130°C. (With the advent of automated processing lines, visual inspection and hard baking are sometimes reversed. The main disadvantage of this reversal is that a problem of any sort with the photoresist process will be caught later on down the line and consequently will take longer to correct.)

ETCH

The early days of silicon processing knew only one type of etching—wet, or liquid, etching. The continued march of technology has seen dry etching begin to be used more frequently. The ideal etchant removes material from regions that are not protected by the photoresist, but does not attack the photoresist. It also does not etch the underlying material, or it etches it very slowly. A measure of this parameter is an etchant's *selectivity*. Selectivity is the ratio of the etch rate in the material being etched to the etch rate of the material that is to "stop" the etchant. A selectivity of 10:1 is considered the minimum acceptable for an easily run manufacturing process.

The ability of an etchant to remove material vertically, but not horizontally, beneath the photoresist layer determines whether it is isotropic (etches the same in all directions) or anisotropic (etches more rapidly in one direction than in another). Figure 4–7 shows the cross sections of two layers of silicon dioxide, one etched with an isotropic, and the other with an anisotropic etchant.

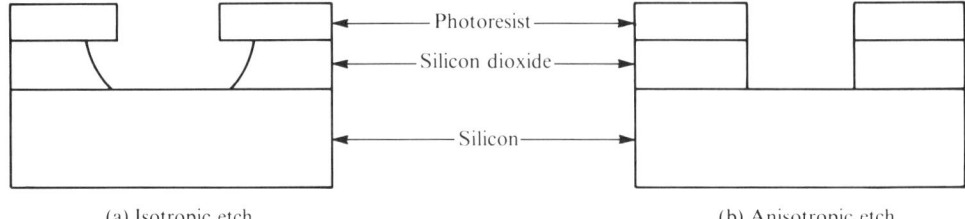

(a) Isotropic etch   (b) Anisotropic etch

Figure 4–7: The cross sections of silicon dioxide regions etched using isotropic and anisotropic etchants.

Liquid etching is accomplished by immersing the wafers in an etching solution at a predetermined temperature. Prior experience and knowledge of the etchant characteristics determines the etch time. If the unprotected area is successfully etched, the wafer is ready to have its resist stripped. Otherwise, the wafer is reimmersed in the etching solution to remove the remaining material. A list of materials commonly encountered in semiconductor processing, together with the chemicals comprising their etch solution, is given in Table 4–2.

The wet etch step is followed immediately by a rinse in deionized water which dilutes the etchant, halting any further removal of material.

Dry etching has become more popular as the demand for smaller linewidths and spaces has increased. The problems of etchant depletion and bubble formation that may be encountered with liquids are not present with dry techniques. In addition, dry etching may be used for highly anisotropic etching. Three general dry-etching techniques have become popular during the last few years.

1. *Plasma etching.* In this technique, wafers masked with photoresist are placed in a chamber which is evacuated. A small amount of reactive gas is allowed back into the chamber. An electromagnetic field is then applied, and the layer that is not protected by the photoresist is etched away by the excited etchant ions. The key to plasma etching is the ability to couple the electromagnetic energy into the reactive species while not heating the rest of the gases in the chamber. Two types of plasma etching systems, barrel reactors and planar reactors, are popular. Figure 4–8 shows a typical barrel reactor, and Figure 4–9 shows a planar reactor.

   A gas (usually a type of Freon) containing fluorine is most often used to etch silicon, silicon dioxide, and silicon nitride. A chlorine-containing gas is used when aluminum is to be etched.

2. *Reactive etching.* This dry-etching technique is a combination of physical and chemical etching. It combines controlled energetic ion bombardment with chemically reactive interaction. Greater anisotropic etch profiles and

Table 4–2: Liquid Etchants for Materials in Semiconductor Processing

MATERIAL	ETCHANT COMPOSITION	TEMPERATURE (°C)	ETCH RATE (Å/MIN)
Thermal SiO_2	Buffered Oxide Etch (BOE) 4:1 to 7:1 NH_4F–HF (49%)	20–30	800–1200
Deposited SiO_2	NH_4F–Acetic acid–H_2O 3:3:2	20–30	1800–2200
Silicon nitride	Phosphoric acid	160–175	50–75
Polycrystalline silicon	HF–HNO_3–H_2O or KOH 1:50:20	20–30	3500–5000
Aluminum	Phosphoric acid–Acetic acid–Nitric acid–H_2O (50:10:2:3)	20–40	2000–6000

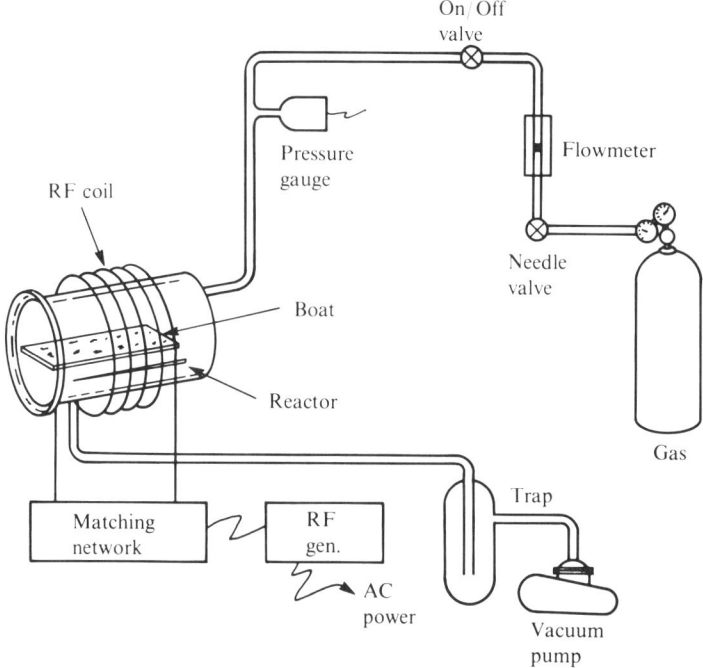

Figure 4–8: A typical barrel reactor.

smaller feature sizes may be obtained using reactive etching. A schematic of a typical etch station is shown in Figure 4–10.

3. *Physical etching* is the use of energetic particles to physically remove material. A beam of charged particles is used in the technique called *ion milling*. Ion milling is similar to reactive ion etching, except that it uses only the energy of motion of the ions to etch material.

STRIP RESIST

Once etching is completed, the photoresist layer should be removed without damaging the rest of the wafer. As with etching, both wet and dry techniques may be used.

1. *Wet.* Negative and positive resists may be removed with a sulfuric acid–hydrogen peroxide or a sulfuric acid–ammonium persulfate solution at temperatures around 100°C at masking steps prior to metallization. Once metal is on the wafer, proprietary formulations available from chemical companies are used. Positive resists are soluble in solvents such as acetone and may be used throughout the process.

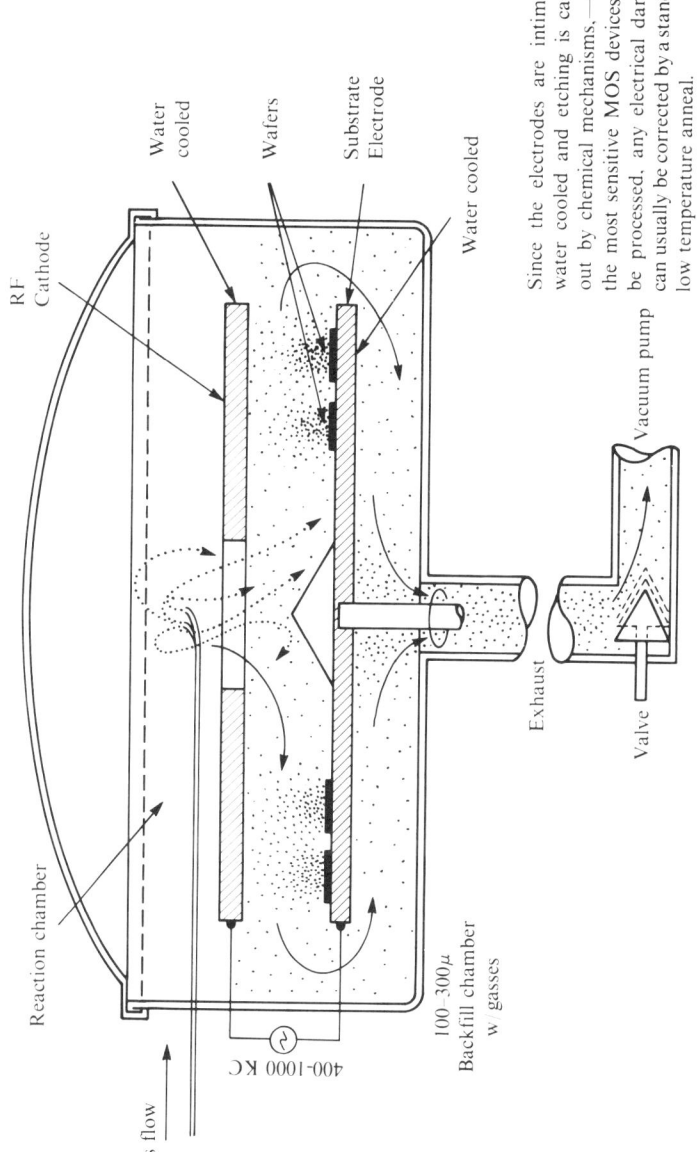

Figure 4–9: A typical planar reactor.

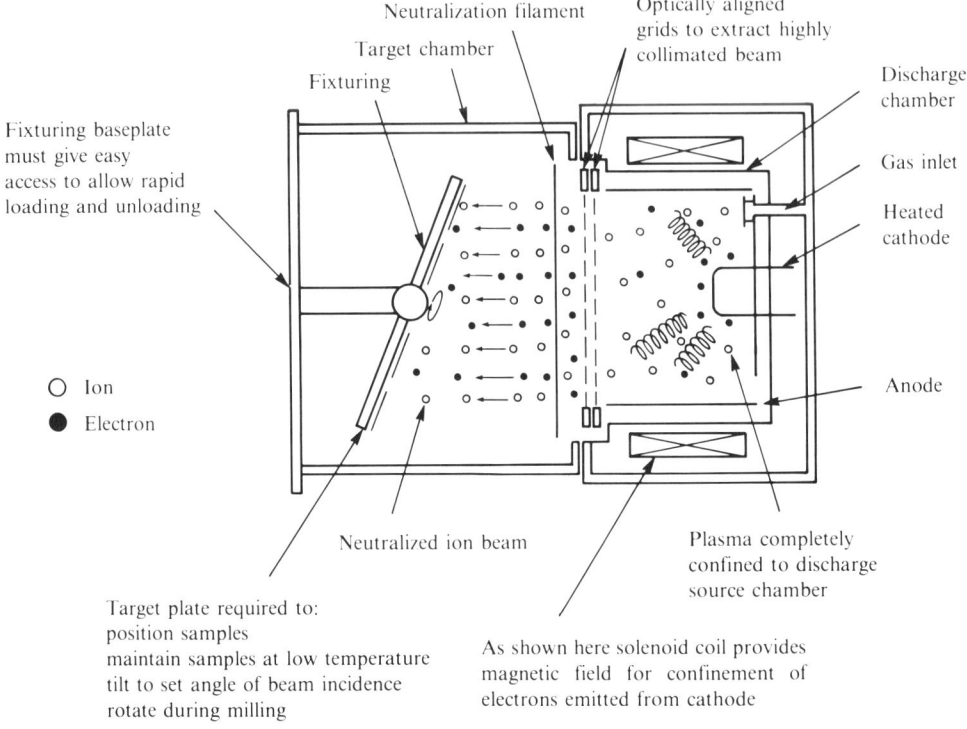

Figure 4-10: A typical reactive etching chamber.

2. *Dry.* Use of oxygen as the resistive gas in a barrel plasma etcher converts the organic molecules of the photoresist to the gases CO_2, N_2, and H_2O, leaving no residue on the wafers.

FINAL INSPECTION

In this operation, each wafer is examined closely using an optical microscope. Wafers that do not meet certain standards are sent back to be reprocessed or are removed from the line. All good wafers are sent on for subsequent processing.

4.4 | Photomasks

The generation of the masks used in the photolithographic process begins with a circuit design developed by one or more engineers. This design may be tested either by building it using actual components or by simulating it on a computer. Extensive evaluation and analysis are performed in either case to determine the circuit's operating characteristics as temperature, supply voltage, and other parameters vary. Once the circuit has been evaluated, translation to the many layers of a mask set begins.

The circuit will eventually be fabricated by sequentially transferring images to the front side of a wafer during the performance of such processes as chemical vapor deposition, epitaxy, predeposition, and drive-in, or during metallization between successive image transfers. Figure 4–11 shows the masks that are transferred to a wafer during a seven-mask bipolar process.

Mask layout is the task of converting the circuit schematic to the final device layout. It consists of the following operations:

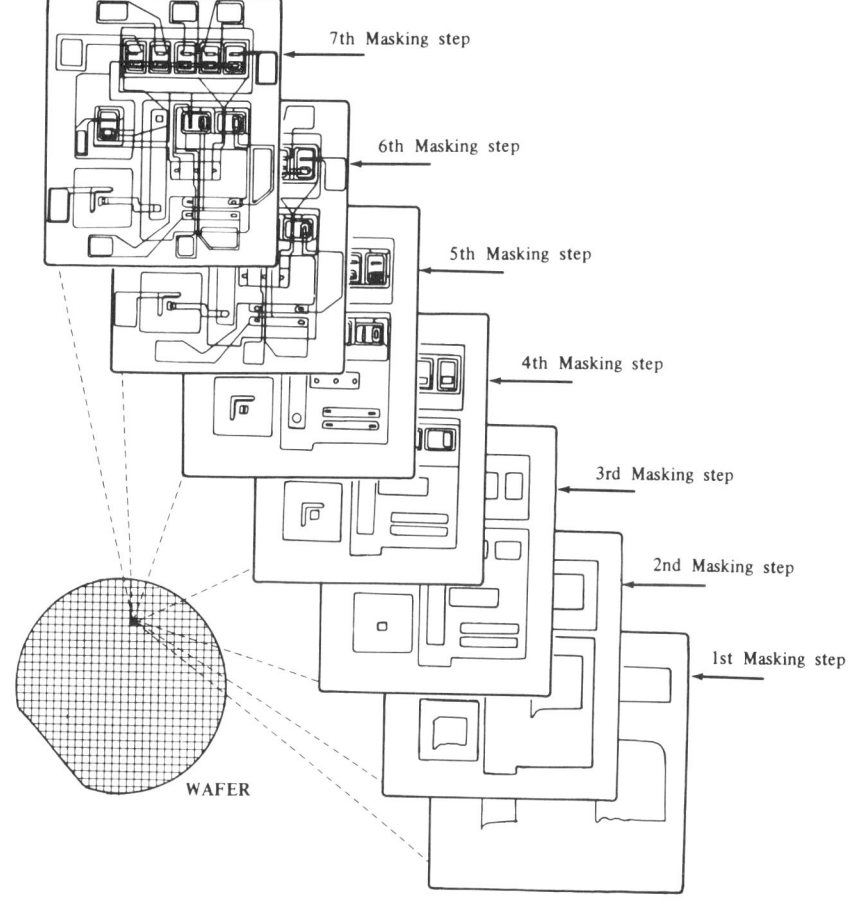

Figure 4–11: The layers transferred to a wafer during a seven-mask bipolar process.

1. Drawing geometrics representing all of the devices in the circuit.
2. Arranging the components of the circuit so that they occupy a minimum of space while making the device interconnection and connection to the outside world as easy as possible.
3. Breaking the composite drawing down into layers for subsequent processing.

The preceding three operations are automated to varying degrees through the use of computer-controlled drawing boards and other equipment, but these aids simply make the circuit easier to design while leaving the creative task of the layout to the operator. After these operations are performed, the actual task of maskmaking begins.

In maskmaking, copies of the layers produced during mask layout are made using computer techniques and are photographically reduced to 10 times the layers' ultimate size. Use of a step-and-repeat camera and the 10× plate results in rows and columns of the identical image being transferred to the glass plate called a "master." A master plate of each layer is produced in this manner. Next, the master is used to manufacture a "submaster," again using a photosensitized glass plate. Finally, many copies or working plates of each submaster are made using more photosensitized glass plates. It is these working plates that are used in the actual transfer of the image to the front side of a semiconductor wafer. (If DSW is to be used for alignment, the 10× or another magnification mask is satisfactory, and the maskmaking process stops.)

Though glass plates covered with photosensitive emulsion are often used in all steps of mask preparation, the emulsion is susceptible to scratches and tears. Thus, alternative materials, including chrome and iron oxide, are sometimes substituted for emulsion. These materials withstand wear better than emulsion masks, but are considerably more expensive. Iron oxide masks have the additional advantage of being transparent to the yellow light used to align

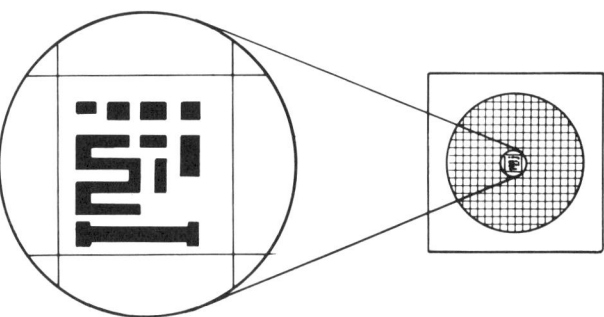

Figure 4–12: A working plate mask.

the masks, while being opaque to the intense ultraviolet light used for the exposure. Visually, each type of mask is a plate of glass with alternate clear and opaque regions as shown in Figure 4–12.

Review Exercises

1. After the development step in a process using positive resist, is the resist left in areas protected by opaque regions of the mask?
2. Using Figure 4–4, determine:
   a. The spin speed needed to obtain a 1.6-μ-thick layer of AZ-1350J resist.
   b. The thickness resulting from a 6000-rpm application using AZ-111 resist.
3. Is a liquid that flows more slowly than another more or less viscous than the first liquid?
4. Name and describe three types of bake oven.
5. List three materials used for the opaque regions of photomasks, and give an advantage for each type.
6. Define photolithography.
7. List and define the four parameters that affect the performance of photoresist.
8. Why is priming necessary in some photoresist processes?
9. What is the most common method of applying photoresist during the manufacture of semiconductor devices?
10. What two parameters are used to control the quality of the finished photoresist layer?
11. What is the purpose of the "develop check" step?
12. Explain the difference between soft baking and hard baking.

5 | Impurity introduction and redistribution

5.0 | Introduction

Impurities are introduced into silicon wafers using two different techniques: thermal predeposition and ion implantation. Regardless of the technique used, the intent of the process is to generate dopant atoms in specific regions of the semiconductor wafer. These doped regions become the transistors, diodes, and other devices that form integrated circuits. This chapter studies the specifics of the technique used to obtain the regions of doped silicon.

5.1 | Definition of Diffusion

Diffusion is the process whereby particles move from regions of higher concentration to regions of lower concentration. The process may be visualized by thinking of a drop of black ink being dropped into a still glass of clear water. Initially, the ink stays in a localized area, appearing as a dark region in the clear water. Gradually, some of the ink moves away from the region of high concentration, and instead of there being a dark region and a clear region, there is a gradation of colors. As time passes, the ink spreads out until it is possible to see through it, though some regions are darker than others. Finally, after a very long time, a steady state is reached and the ink is uniformly distributed in the water. The movement of the ink from the region of high concentration (the ink drop) to the region of low concentration (the rest of the glass of water) is an illustration of the process of diffusion.

Diffusion is illustrated in Figure 5–1. Figure 5–1a shows the distribution of particles at the initial time. As time passes, the distribution moves away

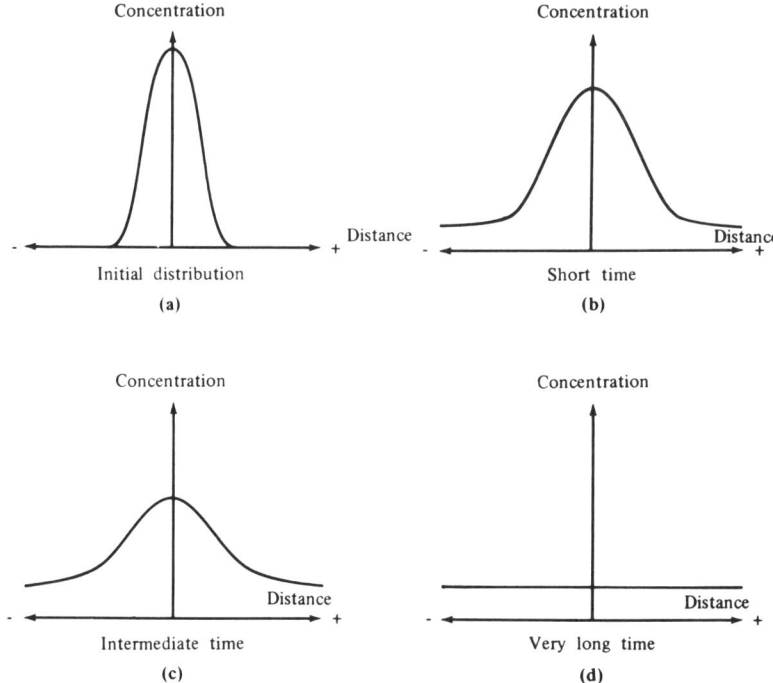

Figure 5–1: Redistribution of impurities with time.

from the center in both directions, as shown in Figures 5–1b and 5–1c. Finally, after a very long time, the particles are uniformly distributed, as shown in Figure 5–1d.

The rate at which the diffusion of particles takes place depends on how fast the particles are moving. This parameter depends, in turn, on the temperature, since hotter particles move faster. The *diffusion coefficient* of a material is the term used to describe this dependence of the rate of diffusion on temperature.

5.2 | The Diffusion Process

Diffusion is used in semiconductor processing to introduce a controlled amount of chosen impurities into selected regions of a semiconductor crystal. The state of the dopant on the surface of the wafer divides the diffusion process into two distinct operations:

1. *Predeposition.* A carefully controlled amount of the desired impurity is introduced into the semiconductor.

2. *Drive-in.* The impurity that has already been introduced into the wafer is redistributed to obtain the final profile.

Each of these operations is considered in great detail in the rest of this chapter.

PREDEPOSITION

During predeposition, the semiconductor substrates are heated to a carefully selected and controlled temperature, and an excess of the desired dopant is made available at the surface of the wafer. (How this excess is obtained is subsequently discussed.) The materials used as dopants enter the crystal structure until a maximum concentration called the *solid solubility* is reached. The solid solubility of one material in another depends on the temperature alone. Figure 5–2 shows the solid solubility of common dopants in silicon.

The solid solubility of a dopant in silicon at a given temperature is the maximum amount of the dopant that will diffuse into the silicon. To obtain controlled conditions during predeposition, an excess of dopant is present at the surface of the wafer. Having more dopant available outside the silicon than can enter the silicon guarantees that solid solubility will be maintained during the predeposition. For example, from Figure 5–2, the solid solubility of phosphorus in silicon at 1000°C is 9×10^{20} atoms/cm^3, while that of boron in silicon at the same temperature is only 2×10^{20} atoms/cm^3.

The substrate temperature determines the solid solubility of the dopant in the semiconductor and, hence, the concentration of the dopant at the surface of the wafer. However, the predeposition time is the other variable needed to completely characterize a predeposition. The predeposition time determines the exact doping profile as one moves away from the surface of the wafer. The effect of ever-increasing predeposition time is shown in Figure 5–3. Figure 5–3a shows the distribution of a dopant in silicon after a small period of time has elapsed. The concentration of the dopant at the surface is the solid solubility determined from Figure 5–2. The dopant concentration quickly decreases as we move away from the surface. Figures 5–3b and 5–3c show the dopant profile for two longer times. The concentration remains the same at the surface of the wafer, but the dopant profile falls off less rapidly as we move away from the surface. Finally, in Figure 5–3d, a very long period of time has elapsed. The dopant profile is now flat throughout the wafer and at the limit determined by the solid solubility.

In the fabrication of silicon devices, the presence or absence of a layer of silicon dioxide on the surface of the wafer determines where the dopant is allowed to enter the silicon. As long as a silicon dioxide layer of sufficient thickness is used, predeposition will introduce dopants only in the desired area. The thickness needed for a particular predeposition can be determined experimentally. Figure 5–4a shows the silicon dioxide thickness necessary to

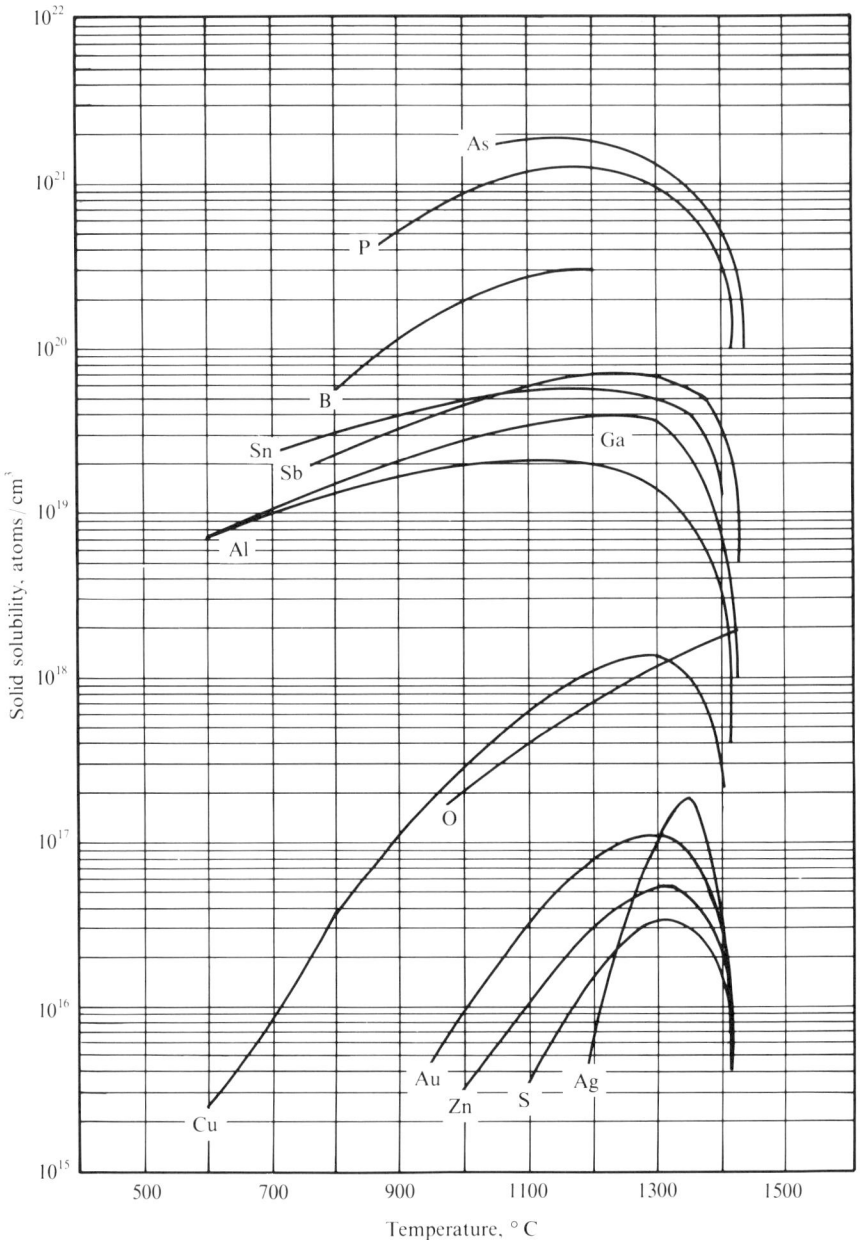

Figure 5–2: The solid solubility of impurities in silicon.

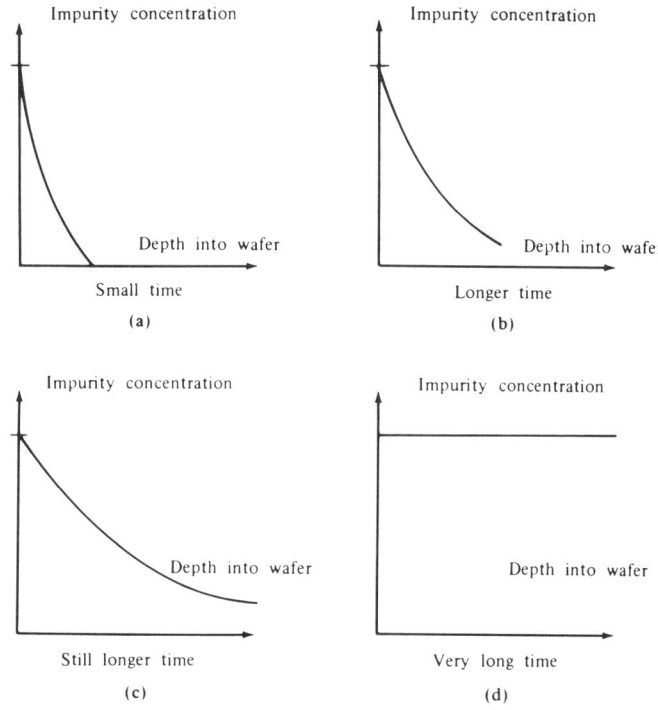

Figure 5-3: Profile of dopant present in a wafer as a function of time, for increasing time.

mask against a boron predeposition, while Figure 5-4b shows the thickness necessary to mask against a phosphorus predeposition.

In wafer fabrication, predepositions are done in furnaces like those used in the oxidation process. Wafers that are ready for predeposition are cleaned to remove any contamination picked up during previous steps. Most predepositions are performed by placing the cleaned wafers in a quartz wafer holder or "boat" and inserting them into a furnace containing an ambient of the desired dopant. The dopant is carried into the furnace by gas flow from the source end of the quartz tube. Enough dopant is deposited on each wafer to guarantee that the solid solubility limit is reached.

The compounds used as dopant sources for this type of predeposition may be solids, liquids, or gases. Solids, often used in powder form, are heated, and a carrier gas is passed over them carrying the dopant into the furnace. The powder may be heated in the end of the furnace itself, or a separate furnace called a source furnace may be used. The arrangement of the source end of a predeposition furnace using a powder is shown in Figure 5-5. The carrier gas used during most predeposition of this nature is nitrogen or a nitrogen–oxygen mixture.

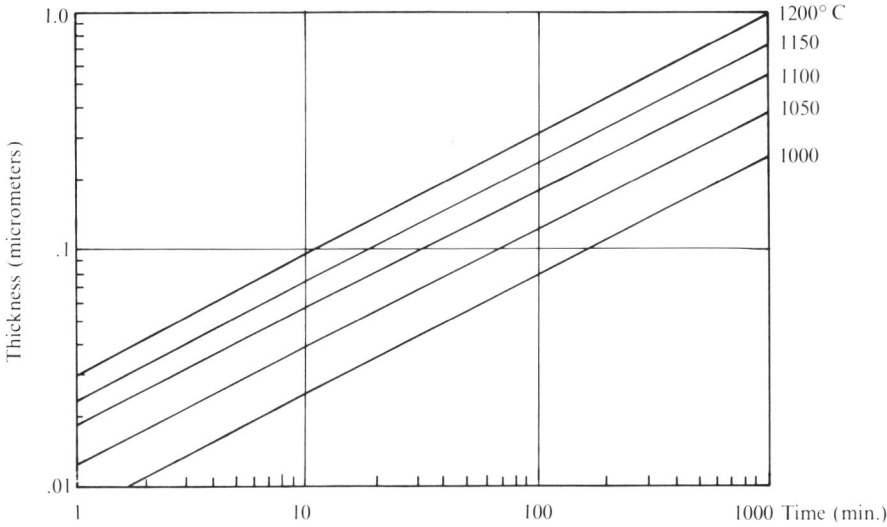

Figure 5–4: (a) Silicon dioxide thickness to mask against boron predeposition.

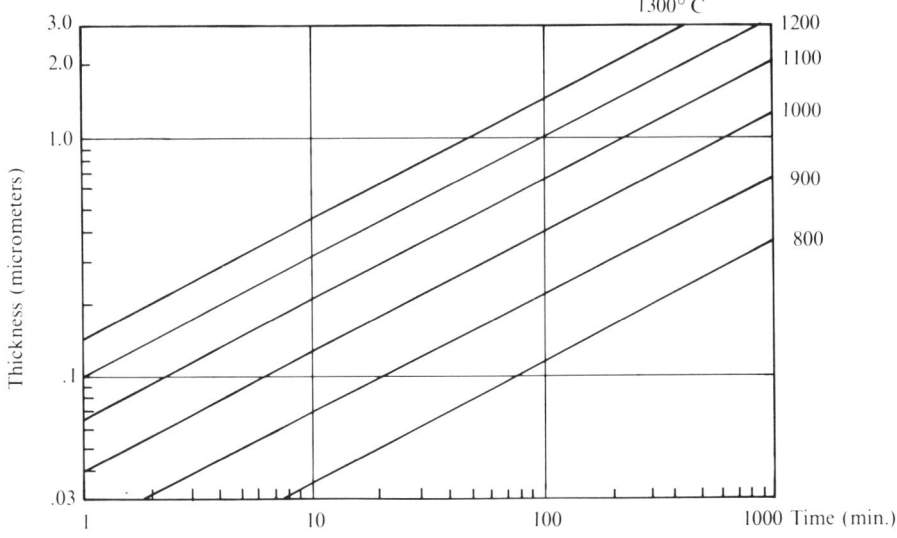

Figure 5–4: (b) Silicon dioxide thickness to mask against phosphorus predeposition.

5.2 The Diffusion Process

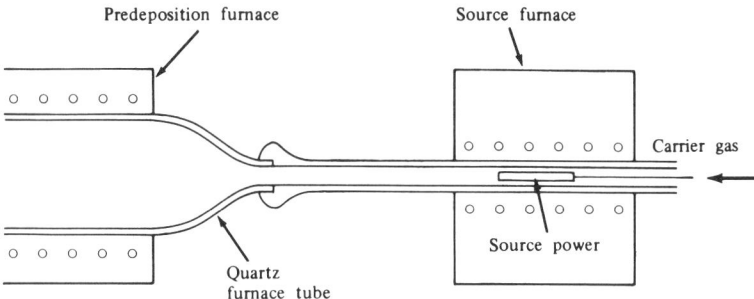

Figure 5–5: Predeposition from a powder using a source furnace.

Experiments have shown that the use of an oxide of the desired dopant produces the best results during predeposition. To guarantee that the dopant reaches the surface of the wafer as an oxide, oxygen is often introduced into the furnace tube along with the dopant.

Liquid dopant sources may be used in a set-up similar to a wet oxidation bubbler. A liquid compound containing the dopant is placed in a bubbler held at a constant temperature. A carrier gas (usually nitrogen) is bubbled through the liquid, whereby it becomes saturated with dopant which it carries into the predeposition furnace. To obtain the oxide of the dopant on the surface of the wafer, oxygen is often introduced into the furnace tube along with the dopant-saturated carrier gas. Figure 5–6 shows a typical liquid-source set-up.

Gaseous diffusion sources are also used in many applications. The use of a gaseous source simplifies the problem of introducing the dopant into the furnace, but there are several other concerns. Some gases used as dopant sources are toxic; care must be taken to ensure that no gas leaks occur. Some dopant gases are chemically unstable and may decompose if improperly stored or if stored too long. Also, the unstable nature of these dopants limits the

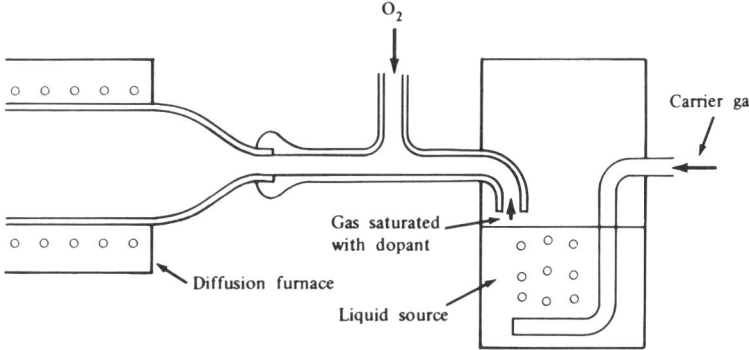

Figure 5–6: Liquid predeposition source.

maximum concentration of the dopant that can be present in the gas. If too low, this maximum dopant concentration may make it impossible to obtain a sufficiently high impurity concentration. The dopant gases must contain the desired impurity, chemically combined with elements that do not adversely affect the predeposition process. In most cases, oxygen is added to the gas flow to guarantee that an oxide of the dopant is deposited on the wafer. A carrier gas like nitrogen may be used to keep the gas-flow velocity along the tube at the proper level. A diagram of a gaseous source system is shown in Figure 5-7.

Another type of predeposition source uses wafers made of a compound of the desired dopant. The source wafers may be prepared by oxidizing them, or they may consist of an oxide of the desired dopant. Both the source and the dopant wafers are placed in a quartz boat and pushed into the furnace. The boat is designed so that the silicon wafers face the source wafers and are a precise distance away from them. Usually, one source wafer has two device wafers facing it, as shown in Figure 5-8.

An ambient gas, usually nitrogen, flows through the predeposition tube. A small amount of oxygen may be introduced to guarantee that the dopant that reaches the surface of the device wafers does so as an oxide. The dopant reaches the device wafers by direct transfer across the gap between the source and device wafers. A summary of solid, liquid, gaseous, and wafer sources is shown in Table 5-1.

A dopant may also be introduced into a semiconductor from a doped layer of silicon dioxide in contact with the surface of the wafer. Two methods are currently used to obtain a doped oxide layer on wafers:

1. A doped layer of oxide may be deposited using a low-temperature chemical vapor deposition process.

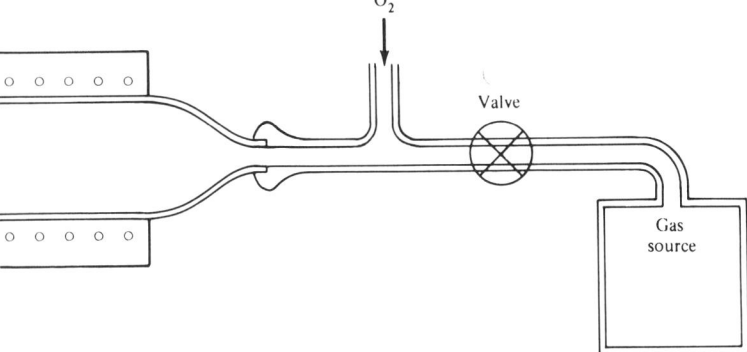

Figure 5-7: Gaseous predeposition source.

5.2 The Diffusion Process

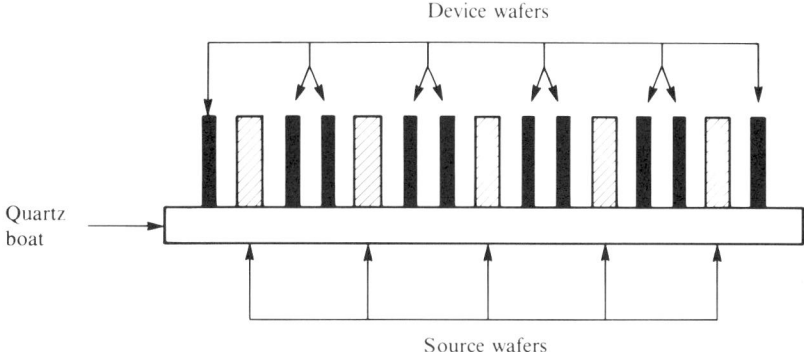

Figure 5-8: Predeposition using a source wafer.

2. A doped layer can be "spun on" using a technique similar to the application of photoresist.

In either case, the wafers are loaded into the furnace with the dopant already on the front side of the wafer. Then, during predeposition, the required amount of dopant diffuses into the semiconductor.

Following predeposition, all surplus dopant is removed from the front side of the wafer by etching. A dilute or buffered solution of hydrofluoric acid is generally used. The wafers are now ready for drive-in.

DRIVE-IN

Drive-in is a diffusion step in which no additional dopant is introduced into the semiconductor. It is done in an oxidizing atmosphere to regrow a protective layer of SiO_2 over the freshly diffused region. During drive-in, the variables of time, temperature, and ambient gases are controlled. These three variables determine:

1. The final junction depth of the diffusion.
2. The final oxide thickness over the newly doped region.
3. The exact profile of the dopant in the semiconductor.

A feel for the relative depth obtained during drive-in is obtained by looking at the "diffusion coefficient" of the particular dopant in silicon. Figure 5-9 shows the root of the diffusion coefficient of some common dopants in silicon as a function of temperature.

Table 5–1: Diffusion Source Chart

TYPE	ELEMENT	SOLID SOLUBILITY (MAXIMUM)	ROOT OF DIFFUSION COEFFICIENT AT 1100°C	SILICON FIT	COMPOUND NAME	FORMULA	STATE	COMMENT (NPN)
	antimony	7×10^{19} (1250°C)	0.110 $\mu m/hr^{\frac{1}{2}}$	O.K.	antimony trioxide	Sb_2O_3	solid	subcollector
	arsenic	1.8×10^{21} (1150°C)	0.090 $\mu m/hr^{\frac{1}{2}}$	good	arsenic trioxide	As_2O_3	solid	closed tube or source furnace; subcollector
n					arsine	AsH_3	gas	subcollector & emitter
					phosphoric pentoxide	P_2O_5	solid	emitters
	phosphorus	1.4×10^{21}	0.329	average	phosphoric oxychloride	$POCl_3$	liquid	emitters
		(1150°C)	$\mu m/hr^{\frac{1}{2}}$		phosphine	PH_3	gas	emitters
					silicon pyrophosphate	SiP_2O_7	solid	wafer source

p	boron	5×10^{20} (1250°C)	0.329 μm/hr	bad	boron trioxide	B_2O_3	solid	base/isolation
					boron tribromide	BBr_3	liquid	base/isolation
					diborane	B_2H_6	gas	base/isolation
					boron trichloride	BCl_3	gas	base/isolation
					boron nitride	BN	solid	wafer source
	gold	10^{14}–10^{17} (800–1100°C)	600 μm/hr	good	gold	Au	solid (evap.)	base-life time control
neither n nor p	iron copper lithium zinc manganese nickel				undesirable impurities from "Contamination"			

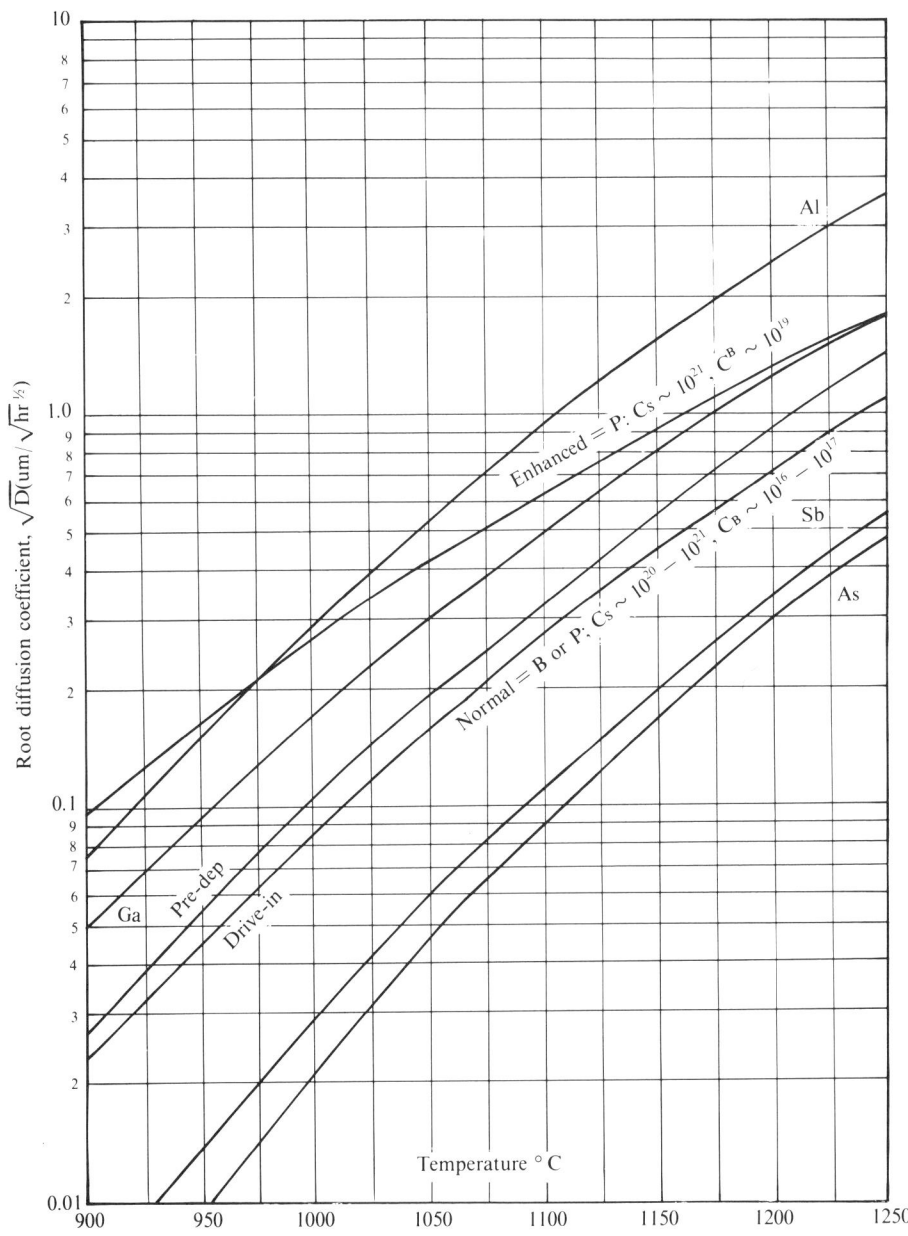

Figure 5–9: Diffusivity of various substitutional impurities in silicon. *(After Trapp, Blanchard, and Shepherd, Ref. 1.)*

5.3 | Diffusion Analysis

The two measurements most frequently used in the evaluation of a predeposition or a drive-in are its sheet resistance and its junction depth. Evaluation of both of these parameters is discussed in Chapter 12, on semiconductor measurements. The only difference between their use in epitaxial evaluation (see pp. 117–119) and diffusion analysis lies in the inability in diffusion analysis to determine the resistivity of a diffused layer from the sheet resistance and the junction depth. The constantly changing concentration as a function of distance from the surface makes the concept of a constant resistivity invalid in diffusion analysis.

5.4 | Mathematics of Diffusion

The mathematics needed to solve for dopant profiles following either a predeposition or a drive-in appears much more difficult than it is. The derivation of the necessary formulas is beyond the scope of this text, but the application of the results to problems is not. The mathematics of diffusion will be studied by looking at predeposition and drive-in separately.

PREDEPOSITION

Predeposition is performed in a high-temperature diffusion furnace with an excess of the desired dopant present at the surface of the wafer. Under these conditions, the concentration C_S of the dopant at the surface of the wafer corresponds to the solid solubility of the dopant at the predeposition temperature, and a uniform and reproducible amount of dopant enters the crystal lattice. The profile of impurities introduced during the predeposition is found using the equation

$$C(x) = C_S \operatorname{erfc} \frac{x}{2\sqrt{D_1 t_1}} \tag{5-1}$$

where

C_S = the solid solubility of the dopant in silicon at the predeposition temperature (from Figure 5–2).

$C(x)$ = the concentration of the dopant at a depth x into the wafer (the complementary error function).

x = the distance from the surface of the wafer.

D_1 = the diffusion coefficient of the dopant at the predeposition temperature (from Figure 5–9).

t_1 = the time the wafers undergo predeposition.

Figure 5–10 is a graph of the normalized complementary error function erfc(z).

The concentration of dopant at some value of x is found in the following manner:

1. Determine the solid solubility C_S of the dopant in the substrate material at the predeposition temperature using Figure 5–2.
2. Determine the root of the diffusion coefficient D_1 of the dopant in the substrate material at the predeposition temperature using Figure 5–9.
3. Evalute the term $2\sqrt{D_1 t_1}$. (Usually, microns are the best units to use.)
4. Using the values of x for which the dopant concentration is being determined, evaluate the term $\dfrac{x}{2\sqrt{D_1 t_1}}$ for each value of x.
5. Each number that results from evaluating the term in step 4 for each value of x corresponds to a point along the horizontal axis of Figure 5–10. Now use the graph to determine

$$\frac{C(x)}{C_S}$$

   for each point.
6. The dopant concentration at a depth x is the product of these graphic values of $C(x)/C_S$ by the solid solubility C_S obtained in step 1.

The complementary error function equation may be rearranged to determine the junction depth resulting from a predeposition if the resistivity or impurity concentration of the substrate is known. The junction depth x_j is the depth at which the dopant concentration $C(x)$ equals the background concentration C_B. Substituting x_j for x and C_B for $C(x)$ in equation 5–1 results in the equation

$$\frac{C_B}{C_S} = \text{erfc}\, \frac{x_j}{2\sqrt{D_1 t_1}} \tag{5-2}$$

The junction depth x_j is determined by the following steps:

1. Evaluate C_B/C_S. This is the value of $C(x)/C_S$ in Figure 5–10.
2. From the figure, determine the value of $x/2\sqrt{D_1 t_1}$ that results in the curve's equaling C_B/C_S.

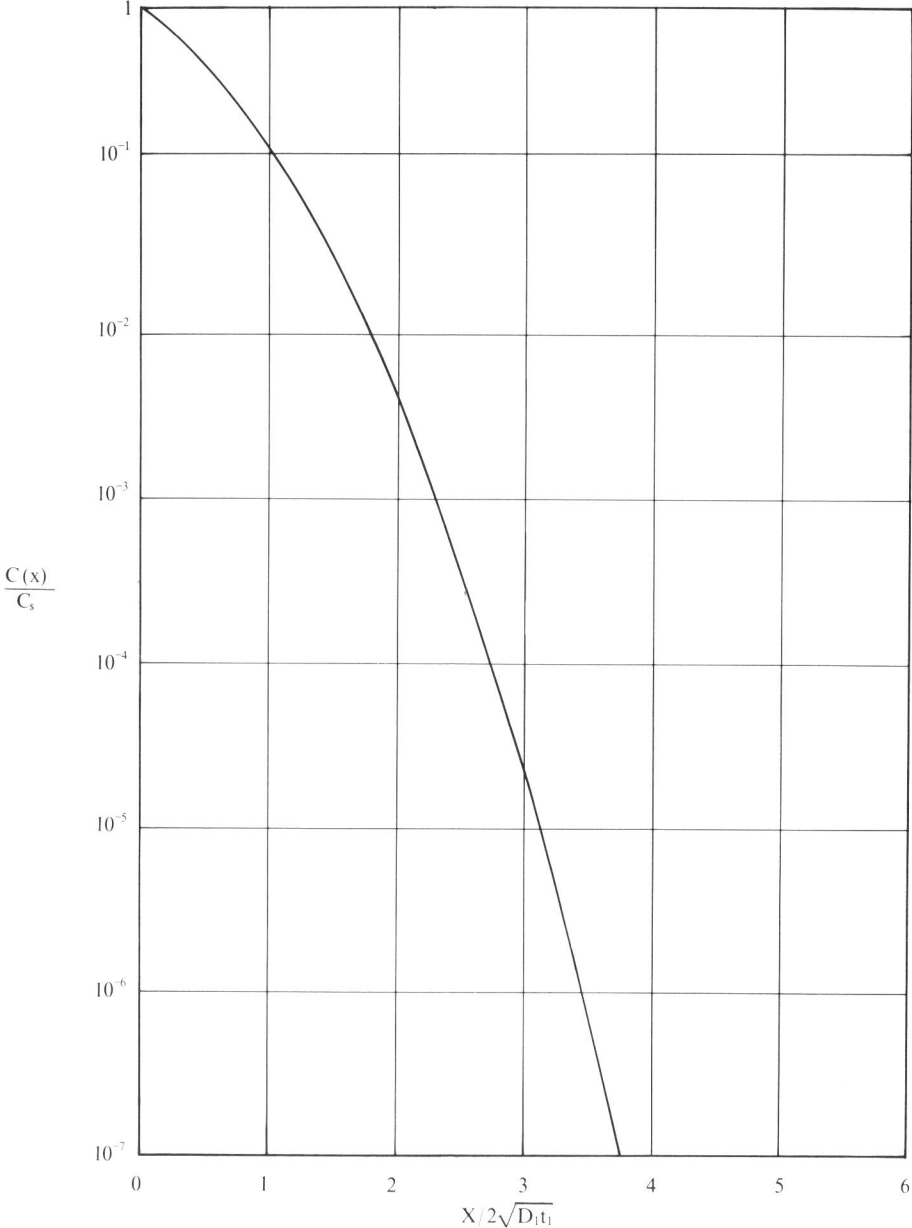

Figure 5–10: Normalized complementary error function.

3. This value must equal $x_j/2\sqrt{D_1 t_1}$. Accordingly, set the two equal to each other, and solve for x_j = (value) $2\sqrt{D_1 t_1}$.

4. Determine x_j, using D_1, the diffusion coefficient of the impurity at the predeposition temperature, and t_1, the predeposition time.

The total amount of impurity Q introduced during a predeposition is found by evaluating the expression

$$Q = C_S \sqrt{\frac{4 D_1 t_1}{\pi}} \text{ (atoms/cm}^2\text{)} \qquad (5\text{--}3)$$

where C_S, D_1, and t_1 are as defined above.

DRIVE-IN

Predeposition has introduced a precise amount of impurity into the crystal lattice, but the junction depth and resulting profile are often not adequate to produce the desired semiconductor devices. Thus, the final junction depth and impurity profile are produced by a drive-in operation. Drive-in is performed in a high-temperature diffusion furnace once the excess dopant remaining on the surface of the wafer from the predeposition has been removed by etching.

When the impurity profile resulting from predeposition is approximated by a rectangle that is very tall and narrow (often called a delta function), drive-in results in a dopant profile expressed by the equation

$$C(x) = \left(\frac{Q}{\sqrt{\pi D_2 t_2}}\right) e^{-x^2/4 D_2 t_2} \qquad (5\text{--}4)$$

where

$C(x)$ = the dopant concentration at a distance x from the surface of the wafer.

Q = the amount of dopant introduced into the crystal during predeposition (equation 5–3).

D_2 = the diffusion of the dopant at the drive-in temperature.

t_2 = the drive-in time.

e = constant = 2.71828.

x = the depth into the wafer.

This equation is valid for most predepositions. Figure 5–11 is a graph of equation 5–4 that is shown as the normalized Gaussian distribution. Note that $C(x)/C_S$ is plotted as a function of $x/2\sqrt{D_2 t_2}$ along the left-hand axis.

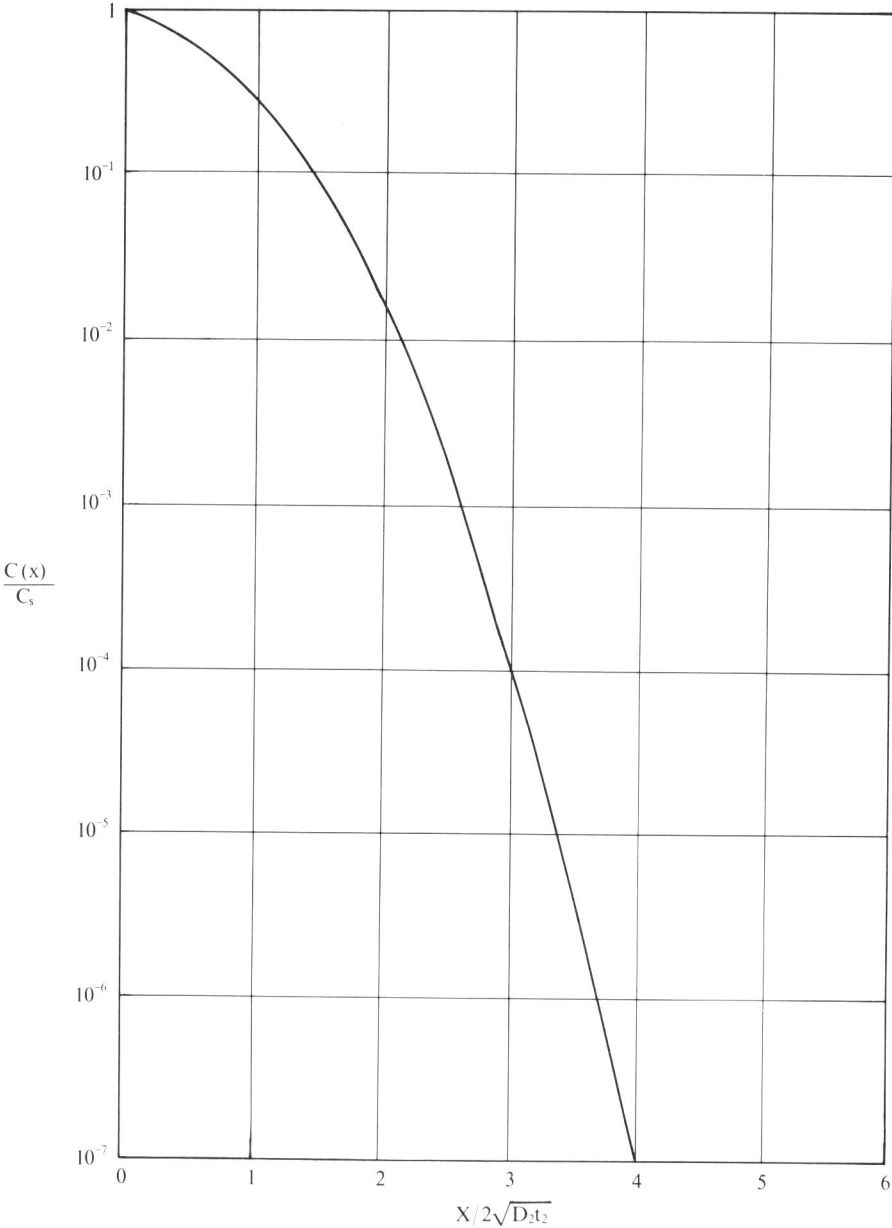

Figure 5–11: Normalized Gaussian distribution.

Figure 5–11 may be used to determine the dopant concentration at a distance x from the surface as follows:

1. Determine Q from the predeposition conditions.
2. Determine D_2 and t_2 from the drive-in conditions.
3. Evaluate

$$\frac{Q}{\sqrt{\pi D_2 t_2}}$$

   and

$$\frac{1}{(4 D_2 t_2)}$$

4. For the chosen values of x, determine $C(x)/C_S$ at each point, using Figure 5–11.
5. Multiply these graphic values by C_S to obtain $C(x)$ at these values of x.

The junction depth following a drive-in is determined graphically as follows:

1. Evaluate C_B/C_S.
2. Using Figure 5–11, find the corresponding value of $x_j/2\sqrt{D_2 t_2}$.
3. Calculate x_j using the value obtained in step 2.

Equation 5–4 may also be used to determine the junction depth resulting from a drive-in if the background concentration C_B of impurities in the substrate is known. As in predeposition, the junction depth x_j is the depth x at which the background concentration equals the dopant concentration. Substituting these values in Equation 5–4, we have

$$C_B = \left(\frac{Q}{\sqrt{\pi D_2 t_2}}\right) e^{-x_j^2/4 D_2 t_2} \tag{5-5}$$

or

$$\frac{C_B \sqrt{\pi D_2 t_2}}{Q} = e^{-x_j^2/4 D_2 t_2}$$

5.4 Mathematics of Diffusion

The value of x_j is determined as follows:

1. Evaluate the expression

$$\frac{C_B\sqrt{\pi D_2 t_2}}{Q}$$

on the left of the equals sign.

2. Determine the exponent w for which

$$e^{-w} = \frac{C_B\sqrt{\pi D_2 t_2}}{Q}$$

3. Set w equal to

$$\frac{x_j^2}{(4D_2 t_2)}$$

and solve for $x_j = \sqrt{4D_2 t_2 w}$.

EXAMPLES

The mathematics just presented may seem like a lot of work, but it is not too difficult if done carefully. To demonstrate the use of this information, let us consider first a predeposition, and then a drive-in, example:

Predeposition

Predeposition is performed on a silicon wafer at 975°C for 30 minutes in the presence of an excess of phosphorus.

1. The concentration of phosphorus as a function of depth is determined as shown below:
   a. From Figure 5-2, $C_S = 8 \times 10^{20}$ atoms/cm^3.
   b. From Figure 5-9, $\sqrt{D_1} = 0.075$ μm/$\sqrt{\text{hr}}$.
   c. $t_1 = 30$ minutes $= 0.5$ hr
      Thus,
      $2\sqrt{D_1 t_1} = 2\sqrt{D_1}\sqrt{t_1} = 2(0.075)\sqrt{0.5} = 0.106$ μm
   Therefore,

$$C(x) = C_S \operatorname{erfc}\left(\frac{x}{2\sqrt{D_1 t_1}}\right) = C_S \operatorname{erfc}\left(\frac{x}{0.106 \text{ μm}}\right) = C_S \operatorname{erfc}(z)$$

Now substitute values for x and determine the associated values for $C(x)/C_S$ at each point. The dopant distribution resulting from this predeposition is determined from Table 5–2 and shown in Figure 5–12.

2. The junction depth is determined by the point of transition from n-type to p-type silicon. If predeposition was done using a .3 Ω-cm p-type wafer, the junction depth is determined either from Figure 5–12 or by solving the proper equation. A .3 Ω-cm p-type substrate corresponds to a background concentration C_B of 10^{17} atoms/cm^3. From Figure 5–12, a C_B of 10^{17} atoms/cm^3 results in a junction depth x_j of about .3 μ. Mathematically,

$$C_B = C_S \, \mathrm{erfc}\left(\frac{x_j}{2\sqrt{D_1 t_1}}\right)$$

where

$$C_B = 10^{17} \text{ atoms/cm}^3.$$

Now,

$$\mathrm{erfc}\left(\frac{x_j}{.106}\right) = C_B/C_S = 10^{17}/8 \times 10^{20} = 1.25 \times 10^{-4}$$

Therefore, from Figure 5–10,

$$\frac{x_j}{.106} = 2.71$$

so

$$x_j = (.106)(2.71) \cong .3 \; \mu$$

Table 5–2: Solution to the Predeposition Problem

x	$\dfrac{x}{2\sqrt{D_1 t_1}}$	$\mathrm{erfc}\left(\dfrac{x}{2\sqrt{D_1 t_1}}\right)$	$C(x)$
0	0	1.00	8.00×10^{20}/cm^3
.1 μ	.943	0.18	1.44×10^{20}/cm^3
.2 μ	1.886	0.75×10^{-2}	6.00×10^{18}/cm^3
.3 μ	2.828	0.63×10^{-4}	5.04×10^{16}/cm^3
.4 μ	3.771	0.97×10^{-7}	7.76×10^{13}/cm^3
.5 μ	4.714	0.27×10^{-10}	2.16×10^{10}/cm^3

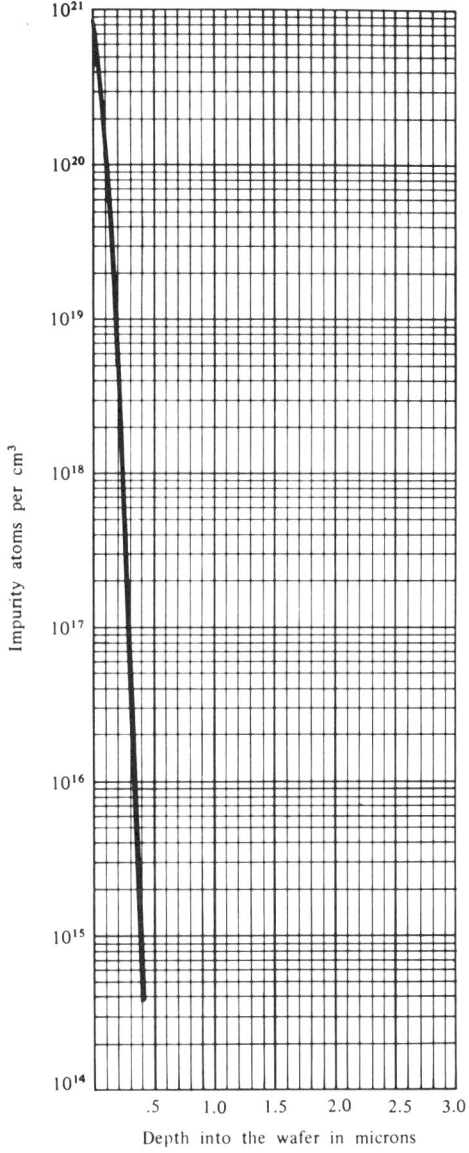

Figure 5-12: Distribution of phosphorus atoms in silicon following the predeposition step.

3. The total amount of dopant incorporated into the wafer per unit area is determined using the equation

$$Q = C_S \sqrt{\frac{4D_1 t_1}{\pi}}$$
$$= (8 \times 10^{20}) \left(\frac{.106 \times 10^{-4}}{\sqrt{\pi}} \right) \text{ atoms/cm}^2$$

Thus, $Q = 4.78 \times 10^{15}$ atoms/cm^2. (Watch the units!)

Drive-in

Using the wafer of part A, determine the dopant concentration as a function of depth following a 50-minute drive-in at 1100°C.

1. The dopant profile is determined as follows:
   a. From Figure 5–9, $\sqrt{D_2} = 0.27$ μm/$\sqrt{\text{hr}}$.
   b. $t_2 = 50$ minutes $= 0.83$ hr; $\sqrt{D_2 t_2} = 0.246$ μm.
   c. Calculate the final surface concentration from

   $$C_S = \frac{Q}{\sqrt{\pi D_2 t_2}} = \frac{4.78 \times 10^{15}}{1.77 \times 0.246} \times 10^4 = 1.1 \times 10^{20}/\text{cm}^3$$

   d. Substitute values for x and determine the associated values for $C(x)/C_S$ at each point. The dopant distribution from this drive-in is determined from Table 5–3 and shown in Figure 5–13.

2. The junction depth of the dopant following the drive-in depends on the resistivity of the initial wafer, but for .3 Ω-cm p-type material as used in the predeposition, Figure 5–13 shows a junction depth of about 1.65 microns.

Table 5–3: Solution to the Drive-in Problem

x	$\dfrac{x}{2\sqrt{D_2 t_2}}$	$\dfrac{C(x)}{C_S}$	$C(x)$
0 μm	0	1.0	1.1×10^{20} cm^3
0.5 μm	1.0	0.36	4×10^{19} cm^3
1.0 μm	2.0	1.8×10^{-2}	2×10^{18} cm^3
1.5 μm	3.0	1.2×10^{-4}	1.3×10^{16} cm^3
2.0 μm	4.0	1×10^{-7}	1.1×10^{13} cm^3

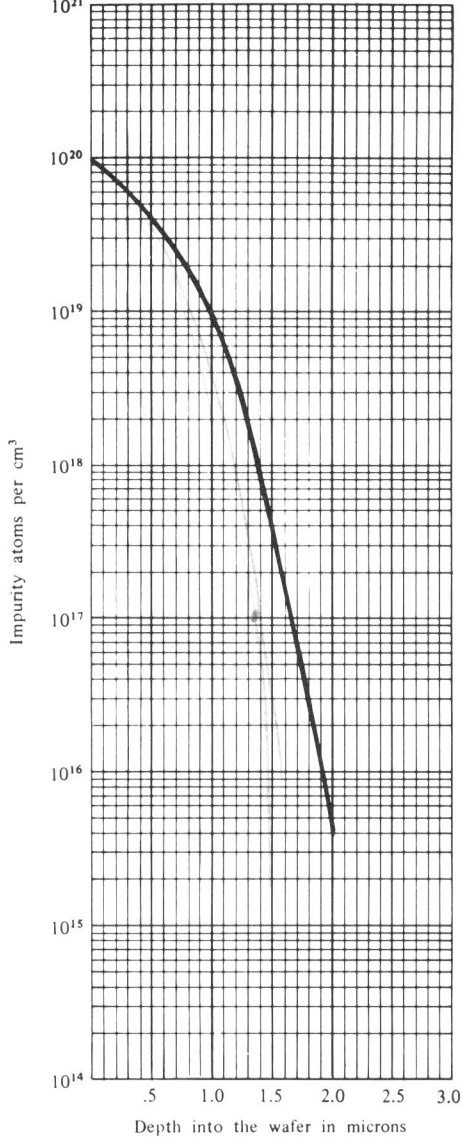

Figure 5–13: Distribution of phosphorus atoms in silicon following the drive-in step.

5.5 | Ion Implantation

A recently developed technique for introducing dopant into semiconductors is *ion implantation*. This process selects ions of a desired dopant, accelerates them using an electric field, and scans them across a wafer to obtain a uniform predeposition. The energy imparted to the dopant ion determines the ion implantation depth.

The first requirement of an ion implantation system is that it generate ions of the desired species. A gaseous source is often used, with ions being generated by boiling them off using a hot filament. The correct ions are separated from any others by bending them through a preset angle using an electromagnetic field. The combination of electric charge and mass determine the amount the various ions change direction. The selected ions are then accelerated using an electric field and strike the target wafer, penetrating the crystal lattice. Figure 5–14 shows the set-up of a typical ion implanter.

Once the doping species has been selected, the two variables that are controlled are the "dose," or the number of ions that reach the wafer per unit of area, and the energy with which they reach the wafer. The dose is controlled by counting the ions as they pass a detector, and the energy is controlled by changing the voltage along the acceleration chamber. The ability to control both dose and energy gives rise to unique applications for this technology.

The regions implanted with the accelerated ions are selected by using either a patterned layer of material such as silicon dioxide or photoresist as a mask. The behavior of various materials as a mask is shown in Figure 5–15. Often, a thin layer of SiO_2 is present over the region of the wafer being implanted, and the ion penetrates this thin layer before entering the underlying substrate. Following implantation and any subsequent cleaning steps, the implanted wafers are often put through a high-temperature furnace to activate any ions that may not have come to rest in electrically active locations in the crystal structure.

Ion implantation offers the ability to precisely control both the amount of dopant introduced in regions of a semiconductor and the depth to which the dopant will travel below the surface. Control of these parameters leads to the applications for which ion implantation is most frequently used, viz., fabrication of high-value or precise resistors, and control of the threshold voltage of field-effect transistors.

5.6 | Mathematics of Ion Implantation

The three most important parameters to control during ion implantation are the dose Q of the dopant and the depth R_P and width ΔR_P of its profile. The

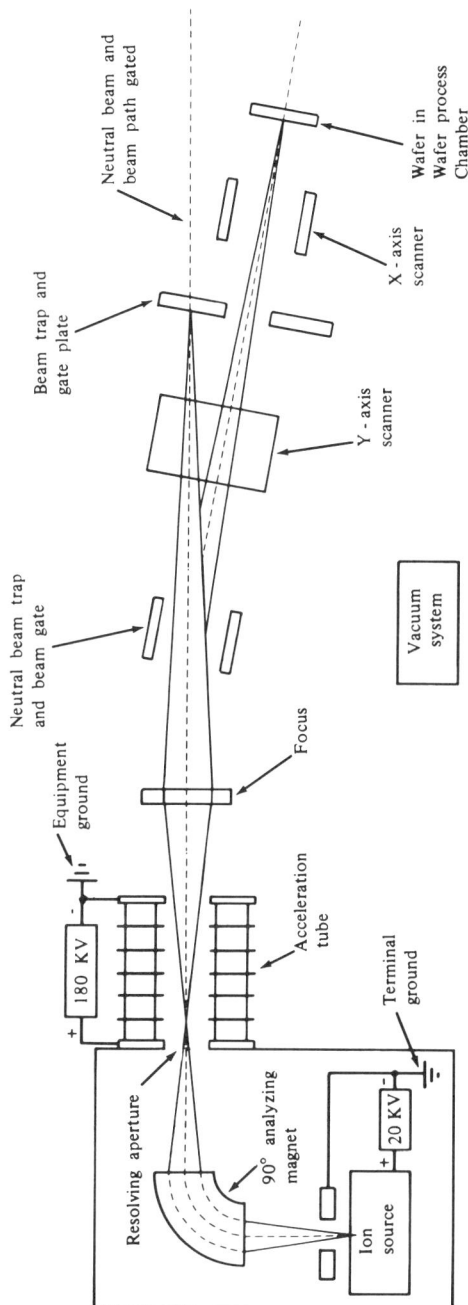

Figure 5–14: Typical ion implantation equipment.

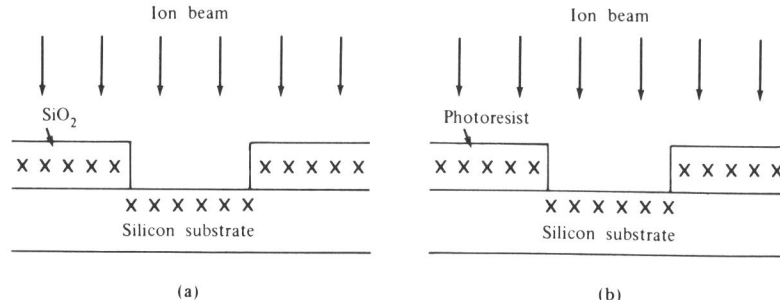

Figure 5–15: Methods of masking during ion implantation. (a) Use of SiO₂ layer as a mask; (b) Use of photoresist as a mask.

dose Q is the number of ions or charges per unit area which are implanted into the wafer surface. By controlling the beam current within the ion implanter, it is possible to maintain an accurate count of the dose, usually in units of total charge per square centimeter.

The range of available doses is wider than that obtainable with thermal predepositions. Typical ion implantation doses range from 10^{11} to 10^{16} ions/cm². Because of this wide range, it is possible to achieve doping profiles unobtainable by any other means. Very accurate control is also possible, and for these reasons, ion implantation is rapidly becoming a standard production tool for a wide array of semiconductor devices and circuits.

Precise control of the ion beam allows accurate control of the depth of penetration and the distribution of the ions being implanted into the semiconductor material. The distance to which the high-energy ions penetrate is a function of the beam energy. The actual distribution of these ions in an amorphous layer is very nearly a Gaussian shape. The mean value of the distribution, R_P, is termed the *projected range* and gives a measure of the average depth of penetration of the ions. This distribution is illustrated in Figure 5–16. The width or spread of the distribution is simply the standard deviation, ΔR_p.

Since the silicon used in the manufacture of semiconductor devices is

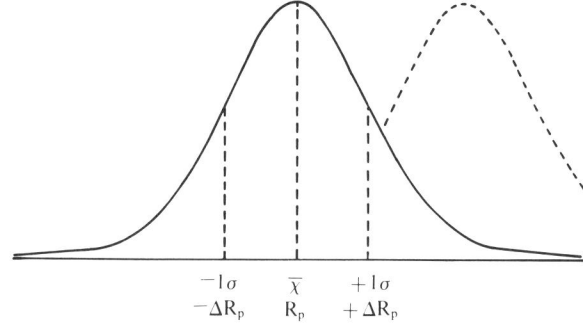

Figure 5–16: Normal Gaussian distribution with channeling.

5.6 Mathematics of Ion Implantation

crystalline, the ion beam may penetrate the silicon lattice on-orientation, i.e., oriented in the same direction as a given crystal face. If on-orientation implantation occurs, some percentage of the ion beam penetrates to a much greater distance by *channeling* between the lattice planes. When this happens, a second Gaussian distribution appears, giving rise to a bimodal distribution as shown in Figure 5–16. Channeling may be minimized by:

1. Tipping the wafer 3–7° off-orientation.
2. Implanting through an amorphous layer such as SiO_2.
3. A combination of techniques 1 and 2.

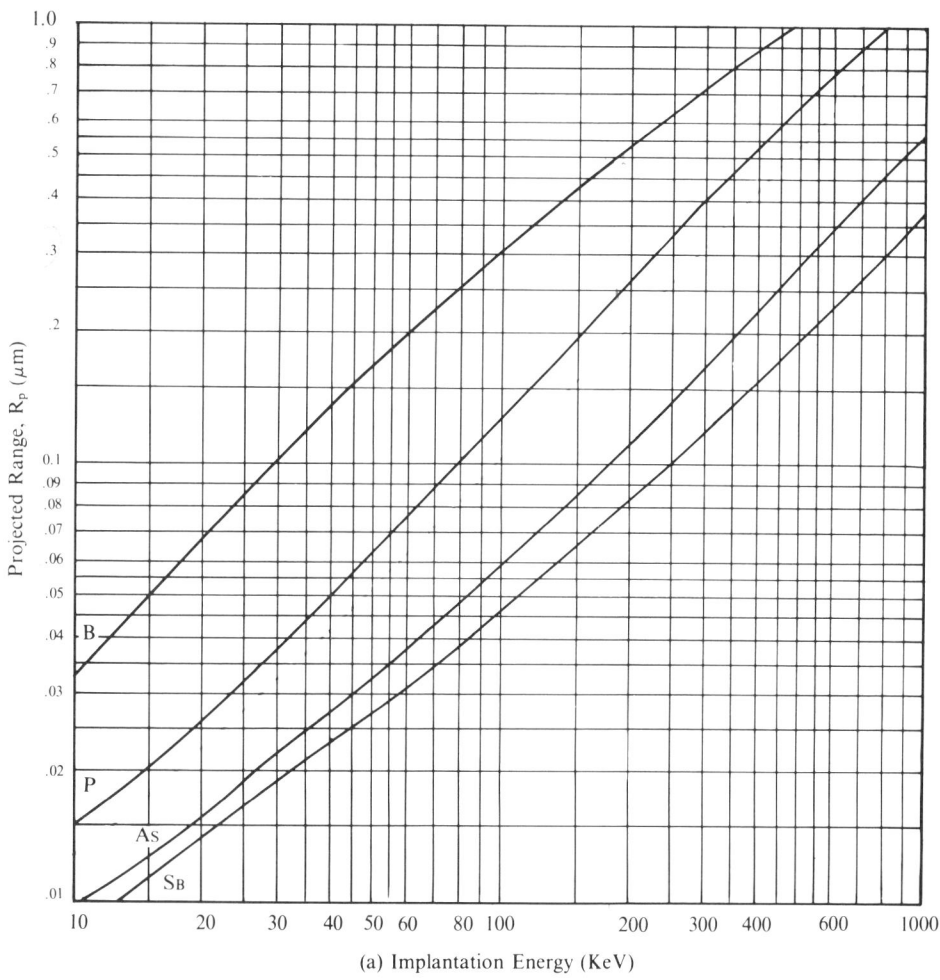

Figure 5–17: (a) Projected range for implantation in silicon. *(After Trapp, Blanchard, and Shepherd, Ref. 2.)*

The Gaussian range statistics R_P and ΔR_P for the major dopants of boron, phosphorus, arsenic, and antimony are given for silicon in Figures 5–17a and 5–17b as a function of energy.

Since ion implantation is a low-temperature process, implantation can take place at room temperature. For this reason, it is unnecessary to use materials, such as thermal oxide, which must be stable at high temperatures. Rather, materials such as metals and photoresists, which are stable at low temperatures, can now be used to mask against or prevent localized implantation. This masking can be done with developed resist which is either chemically removed or plasma stripped if high-dose implants are used. (High-dose im-

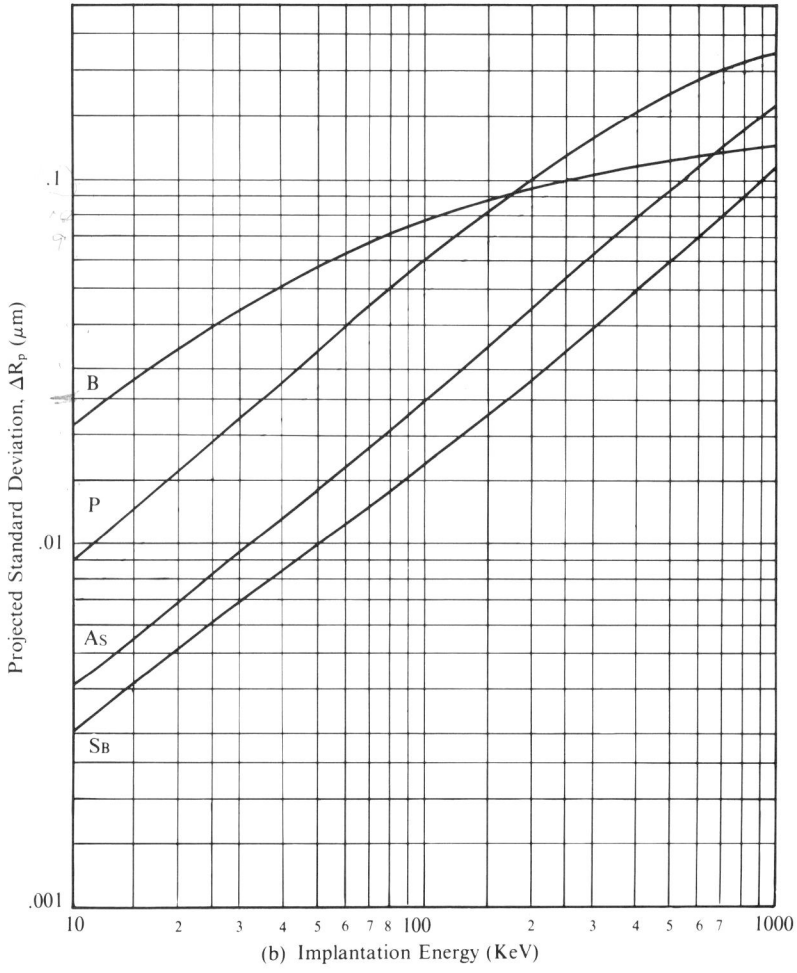

Figure 5–17: (b) Projected standard deviation for implantation into silicon. *(Ref 3.)*

5.6 Mathematics of Ion Implantation

plants chemically alter the properties of the resist, requiring plasma processing to remove it.)

The thickness of the mask material determines the amount of dopant which reaches the substrate through the mask material. A typical processing requirement is that the implant dose be limited by the mask to 0.0001 percent of the ion beam. Figure 5–18 gives the required mask thickness for various dopants and mask materials.

During implantation, the dopant ions may damage or destroy the crystal orientation of the substrate as they move through the lattice. (The amount of damage is a function of implant energy, dose, beam current, and substrate temperature.) In addition, the dopant atoms are electrically inactive since they are not yet part of the silicon crystal structure. Both of these problems are overcome by thermal annealing, which accomplishes two things for low dose implants:

1. The damaged lattice is completely recrystallized above 600°C and proportionally crystallized below 600°C.
2. Most, if not all, of the implanted ions become electrically active as they replace silicon atoms and substitutionally become part of the crystal lattice.

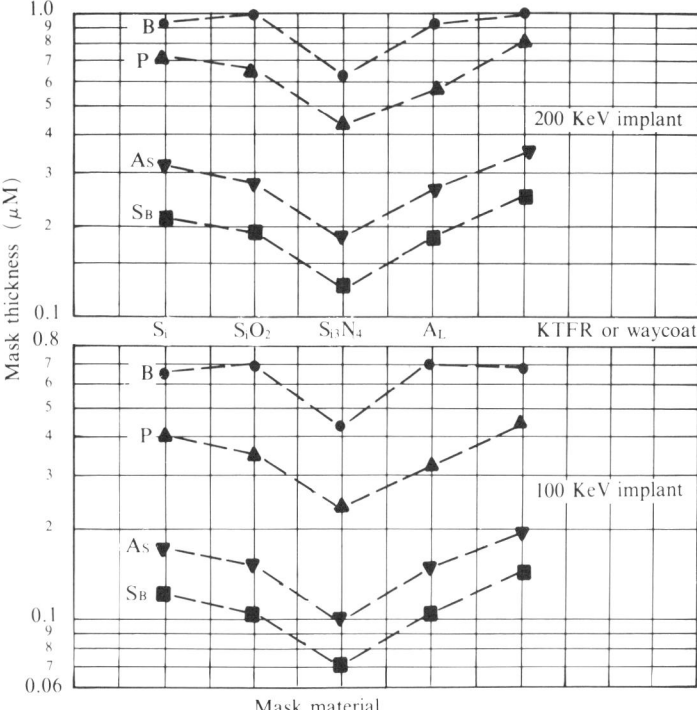

Figure 5–18: Mask thickness for 0.0001% transmission.

5.7 | Limitations of Ion Implantation

The major limitations of ion implantation are dose control and its subsequent measurement. Since there is no outline of the implanted area once the photoresist step is completed, there is no indication as to where or even if an implant has been done.

Actual measurement of the implant is difficult to do, even on test patterns. A more typical method is to use a monitor wafer and either a resistivity or capacitance-voltage measurement. It is also possible to measure changes in reflectivity of special resists on monitor wafers in order to determine dose.

Review Exercises

1. a. Determine the solid solubility of gallium in silicon at 900°C.
   b. Determine the maximum solid solubility of gold in silicon.
2. Does boron or gallium have a higher diffusion coefficient at 1100°C?
3. Determine the diffusion coefficient of phosphorus at 1200°C.
4. What parameter controls the penetration depth of an implanted ion?
5. During predeposition, what parameter determines the concentration of dopant at the surface of a wafer?
6. What two parameters determine the predeposition profile?
7. Determine the oxide thickness necessary to selectively mask a wafer against boron diffusion at 1100°C for 1 hour.
8. List several methods of introducing dopant impurities into a silicon wafer.
9. What three variables determine the depth of the junction during a drive-in diffusion?
10. What two measurements are frequently used to evaluate a diffusion into a silicon wafer?
11. Is it possible to obtain an accurate measure of the resistivity resulting from a diffusion? Explain.
12. Using Figure 5-10, determine:
    a. The erfc of 3.5.
    b. The number whose erfc is 3.5×10^{-3}.

13. Following the predeposition in example A, at what depth would a junction be present if predeposition was done using a p-type substrate doped with:
    a. 5×10^{16} atoms/cm^3.
    b. 5×10^{19} atoms/cm^3.

14. Following the drive-in in example 13, determine the junction depth if drive-in was done using a p-type substrate doped with:
    a. 5×10^{16} atoms/cm^3.
    b. 5×10^{19} atoms/cm^3.

15. What are the impurity profiles after predeposition diffusion? After drive-in diffusion?

16. For a normal diffusion process, does surface resistivity (as measured with a four-point probe) depend directly or inversely on the initial amount Q of the predeposition? Explain.

17. As time progresses during a predeposition, will the surface resistivity increase or decrease? Explain.

18. What are the projected range and standard deviation for:
    a. Boron at 80 Kev in silicon.
    b. Phosphorus at 100 Kev in silicon.

19. What is channeling and how can it be minimized?

20. Arsenic is implanted into silicon at 200 Kev. Determine:
    a. The projected range R_P and standard deviation ΔR_P of the dopant.
    b. The thickness of silicon nitride needed to mask the implant.

REFERENCES (CHAPTER 5)

1. O. D. Trapp, R. A. Blanchard, and W. H. Shepherd, *Semiconductor Technology Handbook*, Technology Associates, Portola Valley, CA, 1981, pp. 4–6.
2. Ibid., pp. 5–9.
3. Ibid., pp. 5–10.

6 | Epitaxial deposition

6.0 | Introduction

Epitaxial deposition is the deposition of a single crystal layer on a substrate (often, but not always, of the same composition as the deposited layer), such that the crystal structure of the layer is an extension of the crystal structure of the substrate. One use of epitaxial deposition in semiconductor processing is in the fabrication of light-emitting diodes. The carefully tailored materials and doping profile that result in the generation of light are usually obtained by using epitaxial techniques in conjunction with high-temperature diffusions. A more frequent use of epitaxial deposition is in the production of both discrete devices and integrated circuits using silicon.

In the fabrication of silicon diodes and transistors, devices with higher switching speed, breakdown voltage, or current-handling capability are obtained using epitaxial deposition. In diode fabrication, a heavily doped silicon substrate used as the starting material results in a diode with lower resistance to current flow. However, high doping concentrations produce lower reverse junction breakdown, so an epitaxial layer of lightly doped silicon of the same conductivity type is deposited on the substrate for the actual fabrication of the junction. The cross section of a typical diode fabricated in this manner is shown in Figure 6–1.

Transistors are fabricated in an analogous manner, using an epitaxial layer for the lightly doped collector region and diffusing in a base and an emitter. Figure 6–2a shows a transistor fabricated this way. The epitaxial layer may also be of the opposite conductivity type, in which case, it serves as the base of the transistor, and the emitter is added during a subsequent high-temperature diffusion. A cross section of a device fabricated in this manner is shown

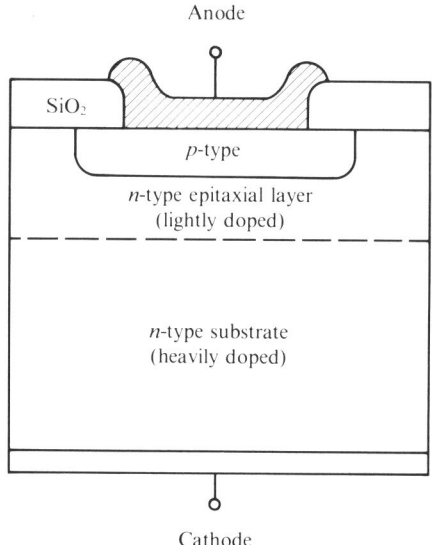

Figure 6–1: Cross section of a diode fabricated using epitaxial silicon.

in Figure 6–2b. In the fabrication of bipolar integrated circuits, a lightly doped silicon substrate of one conductivity type is used as starting material, and a lightly doped epitaxial layer of the opposite conductivity type is deposited. (In most cases, a high concentration of the same type of dopant used in the epitaxial layer is diffused into regions of the substrate prior to epitaxial

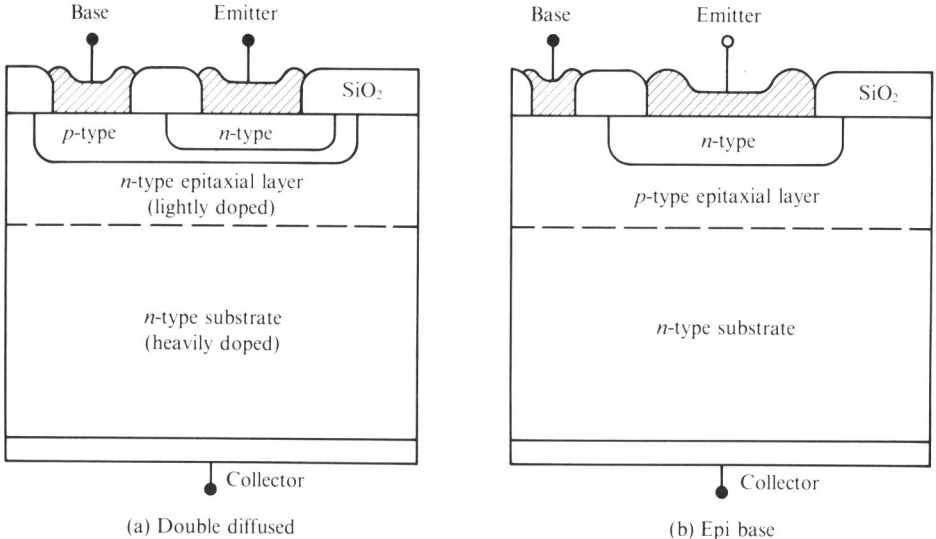

Figure 6–2: Cross sections of transistors fabricated using epitaxial silicon.

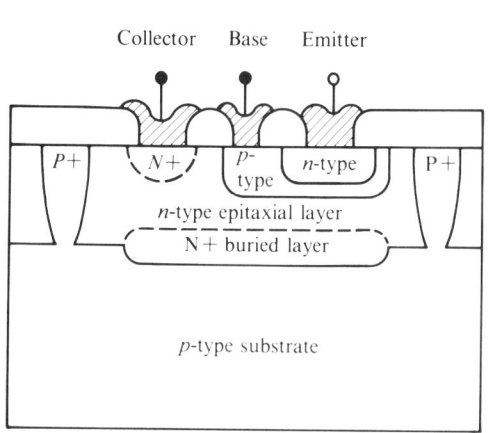
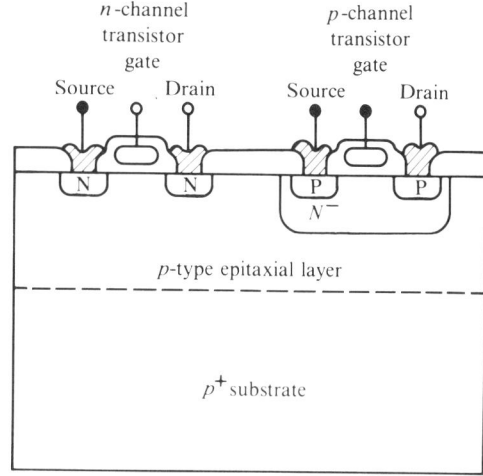

(a) Cross section of a bipolar integrated circuit.   (b) Cross section of a CMOS integrated circuit.

Figure 6–3: The use of an epitaxial layer in conventional integrated circuits. (a) Cross section of a bipolar integrated circuit; (b) Cross section of a CMOS integrated circuit.

deposition to provide a low-resistance path to the active region of devices.) The substrate helps provide electrical isolation between devices in adjacent pockets when the circuit is in operation. A cross section of a transistor in a typical bipolar integrated circuit is shown in Figure 6–3a. MOS integrated circuits—particularly those containing high-voltage devices or complementary MOS (CMOS) circuits—are also often fabricated in a lightly doped epitaxial layer on a more heavily doped substrate, as shown in Figure 6–3b. This chapter discusses both the equipment and the physics of the epitaxial deposition of a silicon layer on a silicon substrate.

6.1 | The Epitaxial Reactor

The deposition of an epitaxial layer of silicon takes place in a specialized piece of equipment called an *epitaxial reactor*. While epi (a shortened version of "epitaxial" often used in the semiconductor industry) reactors are available in a variety of sizes and shapes, they are all combinations of the same subsystems. These subsystems are:

1. *Reaction Chamber.* The volume in which the chemicals react to deposit the epitaxial layer.
2. *Heat source and temperature control.* The section that provides the energy for the chemical reaction.
3. *Gas sources and gas flow controller.* The section that provides the variety of gases that must be supplied for a successful deposition cycle to occur.

4. *Process sequencers and timers.* The sequence of steps in the process of epitaxial deposition must occur in the proper order and for the correct period of time.
5. *Effluent or exhaust control.* The by-products of the reaction process must be dealt with so that they do not harm the environment.

Each of these subsystems is discussed in more detail in the remainder of this section.

REACTION CHAMBER

The reaction chamber provides the carefully controlled environment needed for epitaxial deposition to take place. It is capable of containing the flow of high-temperature gases used throughout the deposition cycle. Epitaxial chambers are made of quartz because of its transparency, ruggedness, and high melting temperature. Three reactor chamber configurations are used in the semiconductor processing industry: horizontal, vertical, and barrel reactors. Their characteristics are as follows.

1. *Horizontal Reactor.* In the horizontal reactor, shown in Figure 6–4, the wafers lie flat on the heated susceptor. The gases enter at one end, flowing horizontally across the wafers where the reaction takes place, and exit at the other end of the chamber. The slight tilt of the susceptor (shown in the figure) compensates for the depletion of the reactive gases as they flow across the wafers. This tilt guarantees that fresh gases are reaching the surface of all wafers.

2. *Vertical Reactor.* In the vertical reactor, shown in Figure 6–5, the wafers also lie horizontally, but the gases enter the chamber in the vertical direction. The reaction chamber is dome shaped, causing the gases to circulate

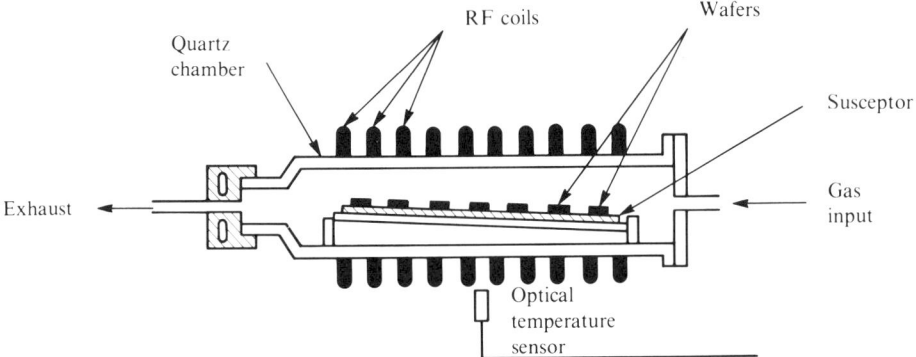

Figure 6–4: Diagram of a horizontal epitaxial reactor.

6.1 The Epitaxial Reactor

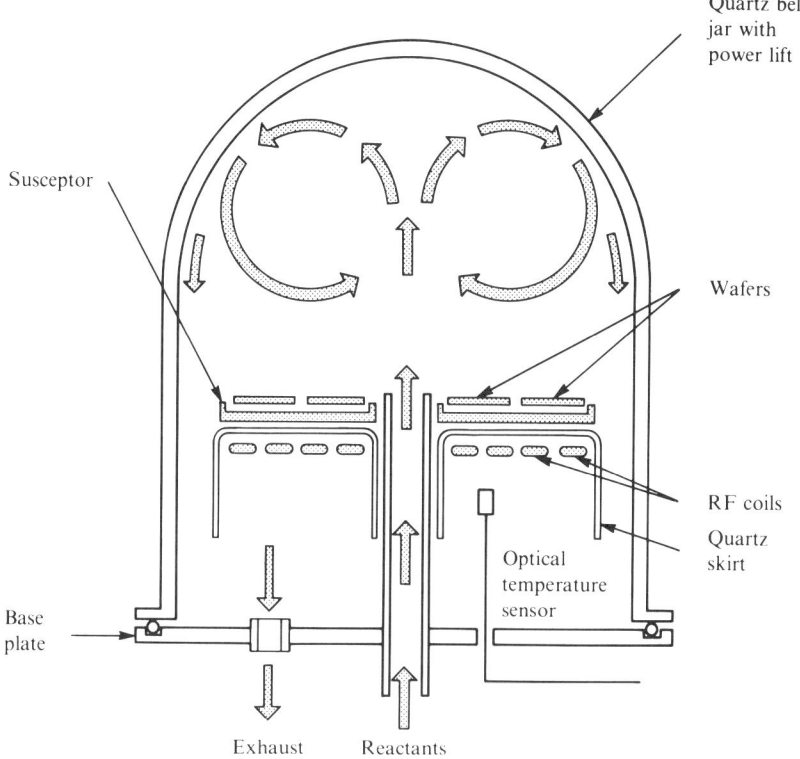

Figure 6–5: Diagram of a vertical epitaxial reactor.

over the wafers (note the arrows in the figure) before they exit the reaction chamber. The susceptor rotates inside the reaction chamber, causing greater mixing of the gases and resulting in more uniform deposition.

3. *Barrel Reactor*. In the barrel reactor, shown in Figure 6–6, the wafers are positioned vertically on the faces of a many-sided susceptor. Gases enter the cylindrical, or barrel-shaped, chamber from the top, reacting as they flow across the surface of the wafers before they exit at the bottom of the chamber. As with the vertical reactor, the susceptor rotates, contributing to a mixing action of the gases and producing uniform deposition.

HEAT SOURCE AND TEMPERATURE CONTROL

In the horizontal and vertical reactors of Figures 6–4 and 6–5, the wafers lie on a wafer holder called a *susceptor*. In addition to supporting the wafers, the susceptor serves as the local source of heat if the reactor is heated using a radio

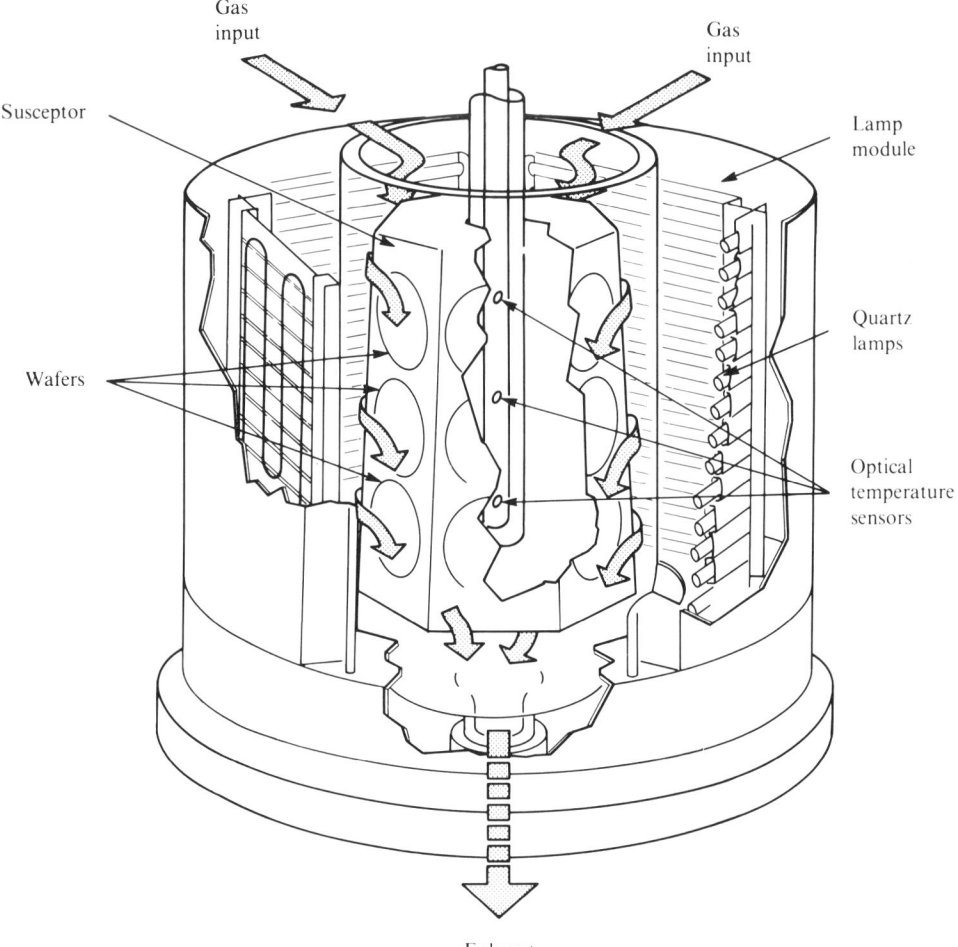

Figure 6–6: Diagram of a barrel epitaxial reactor.

frequency (RF) generator. Radio-frequency- (or induction-) heated reactors inductively couple an electromagnetic field into the susceptor. The RF coils shown in Figures 6–4 and 6–5 act as an antenna, radiating the RF energy. Susceptors usually have a graphite (carbon) body with a thin coating of silicon carbide over the outer surface. The graphite is "susceptible" to being heated by RF energy. The physical behavior is analogous to microwave cooking, where the energy from a microwave source excites water molecules, heating the food. In the epitaxial reactor, the RF energy excites the carbon atoms in the susceptor, heating the susceptor and the wafers that lie on its top surface. The silicon carbide coating prevents the contamination of device wafers with carbon. In an epitaxial process, the deposition proceeds most rapidly on the hottest surfaces in the chamber, which are the front surfaces of the wafers and the susceptor. Deposition on the cooler chamber walls proceeds at a considerably lower rate. (For this reason, epitaxial deposition is called a "cold wall" process.)

The temperature of the susceptor and, hence, of the wafers is monitored using an optical sensing technique. The spectrum of the radiation given off by the susceptor and the wafers is determined by their temperature. A dull red color signifies a much lower temperature than a white color. The optical sensor monitors the color and intensity of the radiation, comparing them to known values to determine the temperature. Techniques that require metallic sensors to be inserted into the RF field do not work because the field will couple into the metal, producing erroneous readings.

The wafers in the barrel reactor of Figure 6–6 are heated using another technique. Ultraviolet energy produced by special quartz lamps heats the wafers through the transparent quartz chamber wall. The temperature is monitored using the optical techniques just described, or with a thermocouple. (A thermocouple may be used because no RF energy is present in a barrel reactor.)

GAS SOURCES AND GAS FLOW CONTROLLERS

A variety of gases flow through the reaction chamber during a typical epitaxial deposition cycle. These gases are categorized as follows:

1. *Nonreactive "flush" gas.* At the start and end of each deposition, a nonreactive gas, usually nitrogen, is used to flush unwanted gases from the reaction chamber. Prior to reactor heat-up, any residual air that might inadvertently have entered the chamber must be removed. During cooldown, following the deposition cycle, any gases remaining from the growth process are flushed out.

2. *Carrier gas.* Before, during, and after the actual growth cycle, the carrier gas maintains uniform flow conditions in the reactor. As the gases responsible for etching, growth, or doping the silicon are added, the flow of the carrier gas maintains a steady-state condition. Hydrogen is most often used as a carrier gas, although helium is sometimes employed. (Nitrogen is not used because it reacts with silicon at the temperatures present during deposition.)

3. *Etchant gases.* Prior to actual epitaxial deposition, etching is performed to remove a thin layer of silicon from the surface of the wafer, together with any crystal damage that is present on it. The etching prepares sites for nucleating or "initiating" the deposition process. Figure 6–7 shows the nucleation sites on the surface of a silicon wafer that have been exposed

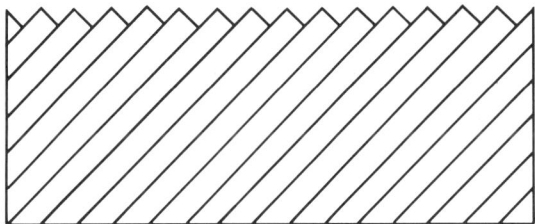

Figure 6-7: The nucleation sites present following the etch step in epitaxial deposition.

after etching in the reactor. HCl is used almost universally as the etchant gas.

The etch rate of silicon as a function of HCl concentration in a horizontal reactor is shown in Figure 6-8. The rate is nearly linear for HCl concentrations of 1-4% in hydrogen. Percentages in this range are often used for etching. However, if the fraction of HCl is too high, a pitted substrate surface results. The maximum allowable fraction of HCl in hydrogen at any temperature is shown in Figure 6-9. A total thickness of 0.25 to 1.0 micron is usually removed from the substrate prior to actual deposition.

4. *Sources of silicon.* The gases used for almost all epitaxial depositions are compounds containing one silicon atom and four other atoms that are

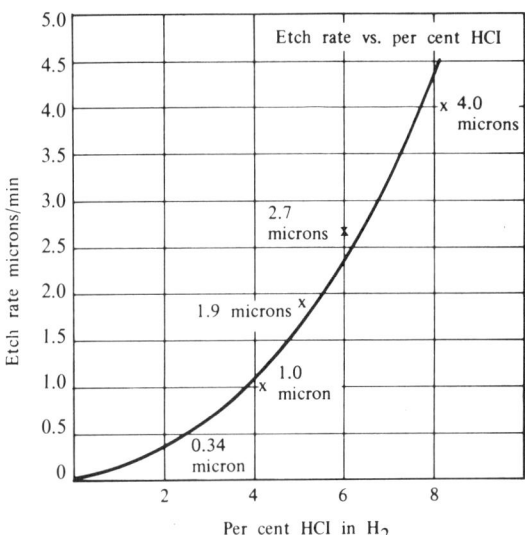

Figure 6-8: Etch rate versus HCl concentration in hydrogen in a horizontal reactor. *(Ref. 1.)*

6.1 The Epitaxial Reactor

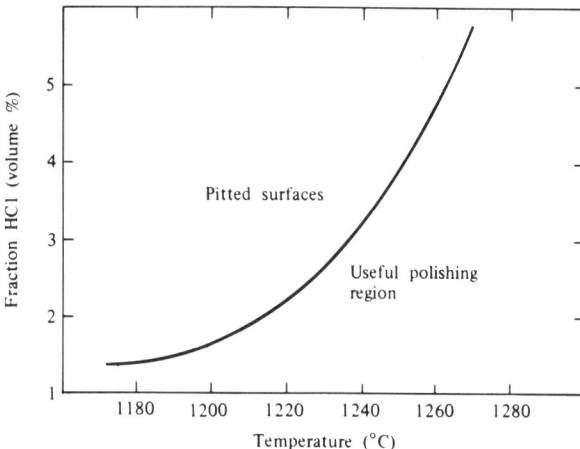

Figure 6–9: Allowable fraction of HCl in H_2 versus temperature. *(Ref. 2.)*

either chlorine or hydrogen, or a combination of the two. Table 6–1 lists these four compounds in order of the range of their deposition temperature. Silicon tetrachloride is a liquid at room temperature, while the other three compounds are gases. Also shown in the table are the chemical formulas and typical deposition rates of each compound.

5. *Dopant gases.* The gases used to control the type of conductivity and the resistivity of the epitaxial layer must be compatible with the gases already present in the reaction chamber. Compounds of hydrogen with arsenic, phosphorus, or boron are the most common dopant gases. Their names and formulas are given in Table 6–2. Typically, these gases are present in small percentages in hydrogen. That they are so diluted allows better control of the amount of dopant in the gas stream because smaller flow variations are required.

The gas flows into the reactor chamber are set using various types of flow meters and flow controllers. Typically, the more sophisticated the method of sensing and controlling the flow, the more expensive the con-

Table 6–1: Silicon Sources, Typical Deposition Parameters

SILICON SOURCE	FORMULA	DEPOSITION TEMPERATURE (°C)	DEPOSITION RATE ($\mu m/min$)
Silane	SiH_4	950–1050	0.1–0.25
Dichlorosilane	SiH_2Cl_2	1025–1100	0.1–1.0
Trichlorosilane	$SiHCl_3$	1100–1175	0.2–2.0
Silicon Tetrachloride	$SiCl_4$	1150–1225	0.2–1.0

Table 6–2: Dopant Gases in Epitaxial Deposition

GAS NAME	CHEMICAL FORMULA	DOPANT TYPE
Arsine	AsH_3	n
Phosphine	PH_3	n
Diborane	B_2H_6	p

troller is. Methods of sensing and controlling the flow include floating a weight (for example, a sphere of material) in a moving gas stream or measuring the ability of a gas stream to cool a heated wire.

PROCESS SEQUENCERS AND TIMERS

Control of the sequence of steps in the process and the time each step begins and ends may be provided in a variety of ways. In the early days of epitaxial deposition, an operator or technician ran the process using his or her memory and a stop watch. The march of technology has seen the sequencer–timer function almost completely automated. On modern equipment, an operator may only have to load and unload wafers and monitor the cycle.

EFFLUENT OR EXHAUST CONTROL

The gases exiting from the reaction chamber are at an extremely high temperature and are often highly reactive. These gases must be diluted, cooled, and made to react in order to eliminate any harmful effects on the environment. This threefold task is accomplished in a piece of equipment called a *scrubber*. In a scrubber, the gases are immediately diluted and cooled by adding large quantities of nitrogen. Next, the gases follow a long route through a continuous spray or shower of water, which further cools them. In addition, any of the gases that are reactive combine with or dissolve in the water. These impurities are then safely dealt with.

6.2 The Epitaxial Growth Sequence

The deposition sequence normally followed in an epitaxial process is as follows:

1. *Substrate Cleaning*. The substrates receive a solvent degreasing operation followed by a series of acid-cleaning steps (H_2SO_4 followed by HNO_3, HCl, and HF is a common sequence) and a drying operation. A physical

scrubbing of the wafer surface may also be included. This cleaning is of great importance since any residual particles may give rise to imperfections in the layer deposited.

2. *Wafer Load.* Following the cleaning sequence, extreme care is taken to ensure that the front side of the wafer is not subsequently touched. The use of vacuum wands on the back side of the wafers is the recommended procedure. Equal care must be taken to guarantee that the cleaned substrates never leave regions bathed with filtered air from laminar flow hoods. While placing the substrates on the wafer holder or susceptor, proper precautions ensure that no particles from the susceptor are transferred to the fronts of the substrates.

3. *Heat-up.* Once the epitaxial system has been sealed, a flow of nitrogen purges any residual gases from the system. Following the purge, the reactor heating system is turned on and heat-up begins. Until a temperature of approximately 500°C is reached, nitrogen is the gas flowing through the system. However, since nitrogen begins to etch silicon at elevated temperatures, hydrogen replaces the nitrogen at higher temperatures.

4. *HCl Etch.* Once the heat-up cycle is completed and the temperature has been verified using an optical pyrometer or other means, removal of a thin region of damaged silicon at the surface of the wafer using an HCl etch (as previously described) takes place. The amount of silicon removed is carefully controlled to guarantee that the characteristics of the devices being fabricated are not adversely affected.

5. *Deposition.* Deposition results in an epitaxial layer with the desired thickness and resistivity. Thickness control is obtained by depositing the layer under growth conditions that minimize the error caused by the slight differences encountered in every run. In epitaxial deposition on <111> silicon, the crystal is 3–7° off axis to allow the easy preparation of nucleation sites by exposing the edges of successive layers of the crystal. Figure 6–10 shows the effect of substrate orientation on the deposition rate of silicon.

Silicon tetrachloride and silane are the two materials that represent the extremes in reaction temperatures. The following discussion examines depositions using these two sources in detail.

HYDROGEN REDUCTION OF SILICON TETRACHLORIDE

Silicon tetrachloride is commercially available in sufficient purity to provide the lightly doped epitaxial layers needed for device fabrication. It is kept in a carefully controlled constant-temperature bath in a liquid state at temperatures near 0°C. Hydrogen flows through or over the $SiCl_4$ to obtain the concentra-

Figure 6-10: Substrate misorientation versus deposition rate. *(Ref. 3.)*

tion of the substance that is necessary for deposition. This concentration is determined by the flow rate of the hydrogen and the temperature of the constant-temperature bath. The effect of temperature on the vapor pressure of $SiCl_4$ is shown in Figure 6–11.

Deposition temperatures of 1150–1225°C are normally used with $SiCl_4$ to obtain a good single-crystal layer. This relatively high temperature may cause significant additional diffusion of already-present doped regions if care is not taken. The reaction commonly recognized as the one resulting in epitaxial growth is

$$SiCl_4\ (g)\ +\ 2H_2\ (g)\ \longrightarrow\ Si\ (s)\ +\ 4HCl\ (g) \qquad (6\text{--}1)$$

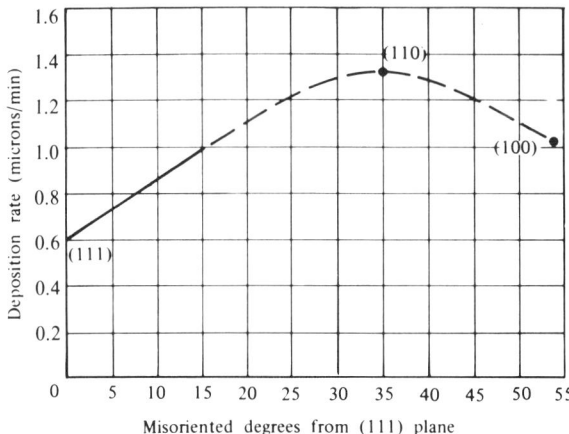

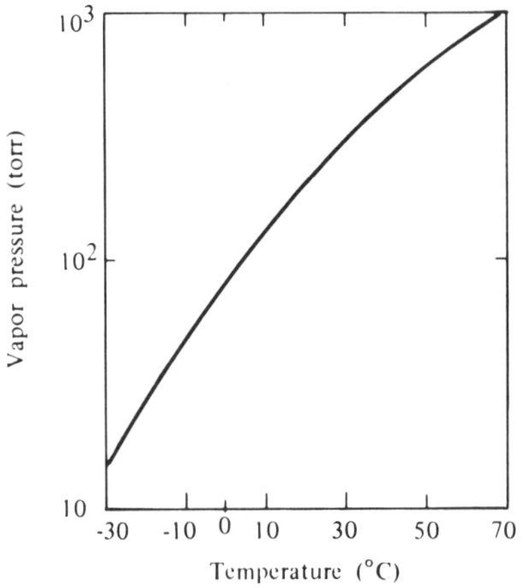

Figure 6-11: Vapor pressure of $SiCl_4$ versus temperature. *(Ref. 4.)*

6.2 The Epitaxial Growth Sequence

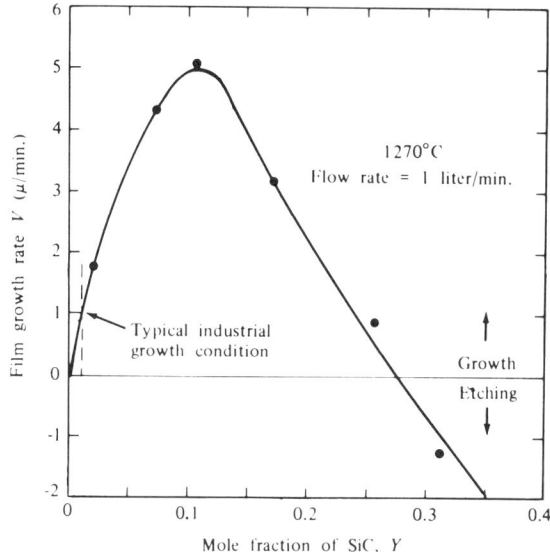

Figure 6–12: Growth rate of silicon versus mole fraction of SiCl$_4$. *(Ref. 4.)*

However, if excessive SiCl$_4$ is introduced, the following competing reaction removes silicon from the substrate:

$$\text{Si (s)} + \text{SiCl}_4 \text{ (g)} \longrightarrow 2\text{SiCl}_2 \text{ (g)} \tag{6-2}$$

The net result of these reactions is shown in Figure 6–12.

PYROLYSIS OF SILANE

Silane is a gas that spontaneously ignites when it comes in contact with air. It is often stored in tanks diluted by hydrogen. The silane or silane–hydrogen mixture is injected directly into the reactor, where the following reaction takes place:

$$\text{SiH}_4 \text{ (g)} \longrightarrow \text{Si (s)} + 2\text{H}_2 \text{ (g)} \tag{6-3}$$

The growth rate as a function of temperature for this reaction is shown in Figure 6–13. Deposition of epitaxial silicon using silane is usually performed in the 950–1050°C temperature range. This deposition temperature range results in less diffusion of previously present heavily doped regions than deposition at higher temperatures.

The desired doping concentration is obtained by adding small concentrations of a dopant gas to the main gas flow. The epitaxial layer doping concen-

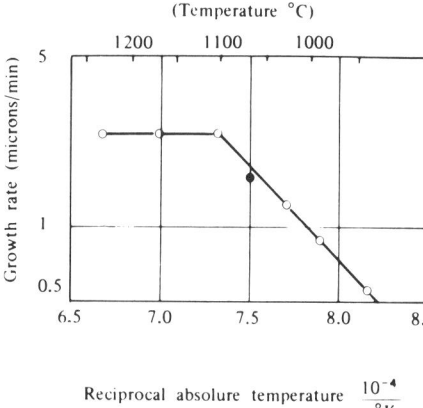

Figure 6–13: Growth rate of silicon versus temperature for SiH_4. *(Ref. 5.)*

tration as a function of the dopant to silicon ratio is shown for phosphorus and boron in Figures 6–14 and 6–15.

6. *Cooldown.* Following the completion of the growth phase of the process, the temperature is reduced while the hydrogen gas flow is maintained. At approximately 500°C, the gas is switched from hydrogen to nitrogen, and the remainder of the cooldown cycle is completed.

7. *Unloading.* The same degree of care must be taken in unloading the sili-

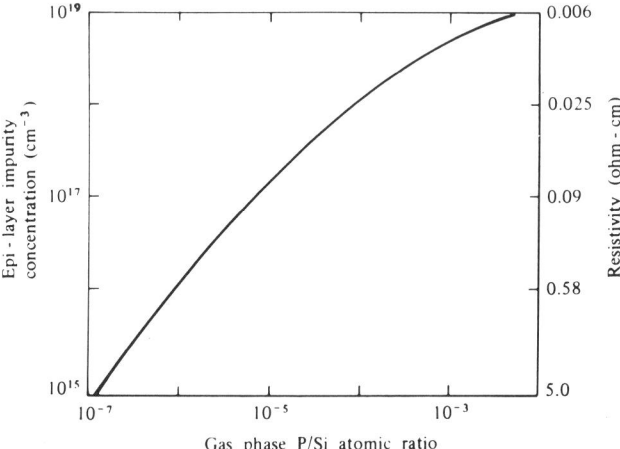

Figure 6–14: Phosphine in gas phase versus phosphorus in epitaxial layer. *(Ref. 6.)*

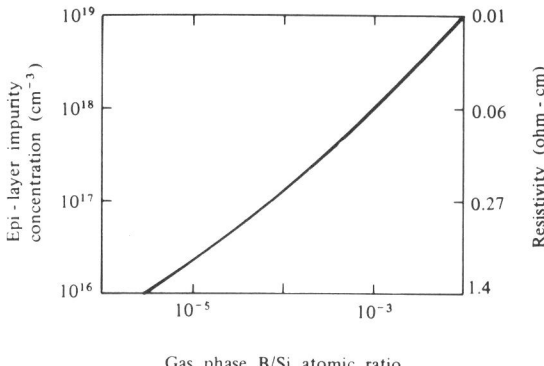

Figure 6–15: Diborane in gas phase versus boron in epitaxial layer. (Ref. 6.)

con wafers as was taken in loading them. The best procedure is to immediately oxidize the wafers to protect their surfaces from possible contamination.

6.3 Evaluation of Epitaxial Layers

The most important parameters in epitaxial deposition are:

1. The thickness of the deposited layer and its variation across the wafer and the run.
2. The concentration of desired impurities and its variation across the wafer and the run.
3. The density and distribution of crystal defects in the deposited layer.

All of these parameters must be within certain bounds for the epitaxial layer to successfully perform its intended purpose.

Epitaxial layer thickness can be measured in a number of ways, three of which are as follows:

1. *Groove and stain (or angle lap and stain).* The epitaxial layer is deposited on a substrate of the opposite conductivity type (p on n or n on p). This epi layer is then grooved or polished beyond its boundary with the substrate. A staining solution is applied to the exposed junction to delineate the junction by darkening either the p-type or the n-type surface. Mono-

chromatic light is then used with a glass cover to generate interference fringes which determine the junction depth, which is related to the number of fringes by the formula

$$d = \frac{n\lambda}{2}$$

where

d = the junction depth

n = the number of interference fringes

λ = the wavelength of the monochromatic light.

2. *Etch pit depth.* The presence of defects at the interface between the substrate and the epitaxial layer generates crystal imperfections that propagate to the surface of the wafer along the crystal planes. A preferential etch is used to delineate the etch pits. The thickness d of the epitaxial layer is geometrically related to the length a of the sides of the exposed etch pits. For <111> silicon, shown in Figure 6–16, $d = 0.816a$.

3. *Infrared interference.* The boundary between the substrate and the epitaxial layer represents an interface that reflects radiation of the correct wavelength. By determining the wavelength of the radiation that produces certain interference characteristics, the thickness is determined as shown in Figure 6–17.

The impurity concentration is determined using various techniques, all of which, unfortunately, are rather time-consuming and tedious. Some of these are:

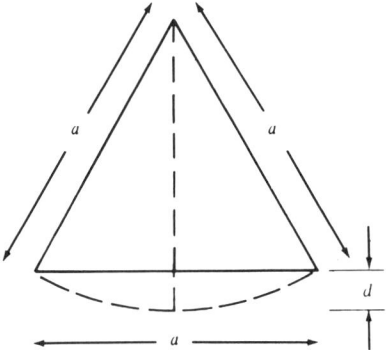

Figure 6–16: Etch pit determination of epitaxial layer thickness.

6.3 Evaluation of Epitaxial Layers

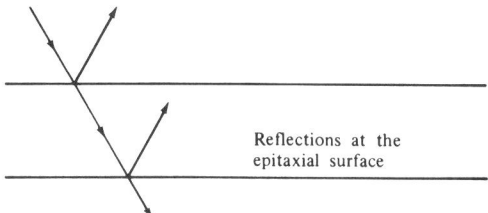

Figure 6–17: Use of infrared interference to determine epitaxial layer thickness.

1. *Sectioning.* Layers of silicon are removed from the surface of the wafer using anodic oxidation or etching, and the sheet resistance of the newly exposed surface is determined. The data can be mathematically manipulated to give the impurity concentration.

2. *Reverse-bias C-V technique.* A layer of metal is deposited on the surface of the wafer to form a Schottky barrier diode. The diode is reverse biased, and the capacitance as a function of reverse-biased voltage (C-V) is determined. The information is then mathematically manipulated to give the impurity concentration.

3. *Lap and spreading resistance probe.* The profile to be measured is exposed using a lapping technique, and fine probes are used to determine the resistivity of the material at positions along the exposed surface of the profile, as shown in Figure 6–18. This technique is often referred to as "SRP." The data are then reduced, resulting in a plot of the dopant profile as a function of depth into the wafer.

The quality of the epitaxial layer is measured using an etching technique that preferentially exposes defects in the crystal structure. The number, location, and kind of defects determine the crystal quality of the epitaxial layer, as was discussed in Chapter 2.

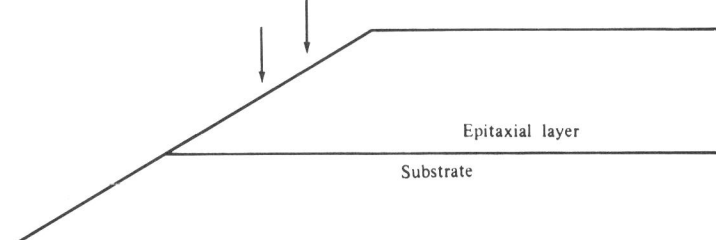

Figure 6–18: Lap-and-spreading resistance probe (SRP) technique for determining impurity concentration.

Review Exercises

1. Must an epitaxial layer be deposited on a substrate of the same composition?
2. Using Figure 6–10, determine the misalignment angle that will result in the highest deposition rate.
3. What is the maximum percentage of HCl that could be used at 1250°C without resulting in a pitted surface?
4. a. Determine the mole fraction of $SiCl_4$ resulting in a maximum film growth rate?
   b. Why isn't epitaxial silicon grown under these conditions?
5. How are nucleation sites created in a silicon wafer?
6. Write the reaction equations and give a brief explanation for the two most widely used industrial epitaxial deposition techniques.
7. Determine the thickness of the resulting epitaxial layer using silane at 1050°C for 5 minutes.
8. Name and discuss two methods of heating wafers in an epitaxial reactor.
9. Why is epitaxial deposition done using a cold-wall reactor?
10. What are three important parameters in epitaxial deposition?
11. The wavelength of light used to determine a junction depth is 0.3 μ. If eight fringes are present, what is the epitaxial layer thickness?
12. List and describe two methods of determining the thickness of an epitaxial layer.
13. Determine the thickness for <111> silicon of an epitaxial layer in which the sides of etch pits are 1.838 μm.

REFERENCES (CHAPTER 6)

1. K. E. Bean and P. S. Gleim, "Vapor Etching Prior to Epitaxial Deposition of Silicon," paper presented at the Fall meeting of the ECS, 1963.
2. G. A. Lang and T. Stavish, "Chemical Polishing of Silicon with Anhydrous Hydrogen Chloride," *R. C. A. Review 24* (December 1963), pp. 488–498.
3. S. K. Tung, "The Effects of Substrate Orientation on Epitaxial Growth," *J. Electrochem. Soc. 112* (April 1965), pp. 436–438.

4. H. C. Theuren, "Epitaxial Silicon Films by the Hydrogen Reduction of SiCl$_4$," *J. Electrochem Soc. 108* (July 1961), pp. 649–653.

5. B. A. Joyce and R. R. Bradley, "Epitaxial Growth of Silicon from the Pyrolysis of Monosilane on Silicon Substrates," *J. Electrochem Soc. 110* (December 1963), pp. 1235–1240.

6. R. M. Warner, Jr., ed., Motorola, Inc. Semiconductor Products Division, *Integrated Circuits* (New York: McGraw-Hill), 1965.

7 | Nonepitaxial chemical vapor deposition

7.0 | Introduction

Chemical vapor deposition (CVD) is the formation of a stable compound on a heated substrate by the thermal reaction or decomposition of gaseous compounds. Epitaxial growth is a highly specific type of CVD that requires that the crystal structure of the substrate be continued through the deposited layer. For this reason, it was covered separately in Chapter 6. In this chapter, nonepitaxial CVD and its applications will be covered.

CVD may be accomplished in many ways, but all types of CVD equipment need to have the following basic sections:

1. Reaction chamber
2. Gas control section
3. Timing and sequence control
4. Heat source for substrates
5. Effluent handling

The variety of ways of implementing each of these sections leads to a great number of individual reactor configurations.

The purpose of the reaction chamber is to provide a controlled environment for the safe deposition of stable compounds. The chamber boundary may be quartz, stainless steel, aluminum, or even a blanket of nitrogen. Reaction

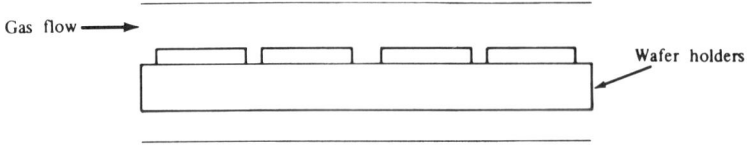

Figure 7–1: A horizontal reaction chamber.

chambers are of the following four general types, depending primarily upon gas flow:

1. *Horizontal systems.* Wafers are placed horizontally on a wafer holder (boat or susceptor), as shown in Figure 7–1. In these systems, the gas flows in one end of the tube, across the wafers, and out the other end.
2. *Vertical systems.* Wafers are placed on a susceptor with the gas flow incident to the wafers from the top, as shown in Figure 7–2. The susceptor usually rotates to produce uniform temperatures.
3. *Cylindrical, or barrel, systems.* Wafers are placed vertically on the outer surface (or sometimes the inner surface) of a cylinder. Gases flow into the chamber from the sides, and the susceptor usually rotates. Such a chamber is shown in Figure 7–3.
4. *Gas-blanketed downflow system.* Gases flow downward, as in a vertical system, while wafers are on a moving wafer holder, as in a horizontal system.

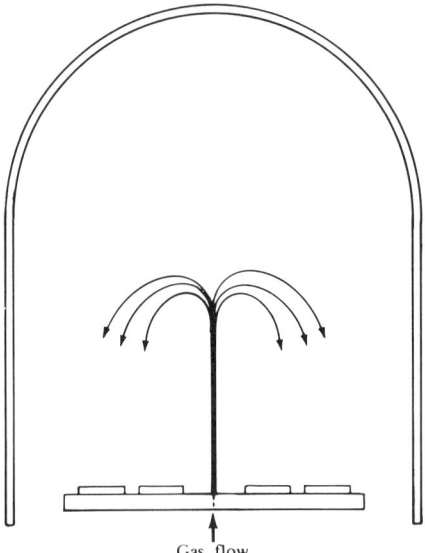

Figure 7–2: A vertical reaction chamber.

7.0 Introduction

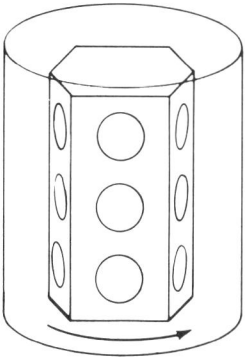

Figure 7–3: A cylindrical or barrel system.

A blanket of inert gas (usually nitrogen) keeps the reaction species separate from the outside atmosphere, as shown in Figure 7–4.

The gas flow control section determines the types and amounts of gas that flow into the reaction chamber. The exact type of flow controller used depends on the accuracy needed in the particular application. In general, the greater the percentage of control needed, the more important (and expensive) the flow controller is.

The timing and sequence control section is responsible for the overall running of the CVD equipment. It may vary in complexity from manual on–off buttons controlled by an operator to a completely computer-controlled automatic programmer.

Heating sources are divided into two general categories: cold-wall systems and hot-wall systems. This distinction is made because, in cold-wall systems, reactions that lead to deposition on the chamber walls proceed at a relatively slow rate. In hot-wall systems, the deposition process will take place on the reaction chamber walls as fast as or faster than on the wafers and the susceptor. Heating in a cold-wall CVD system is accomplished through the use of

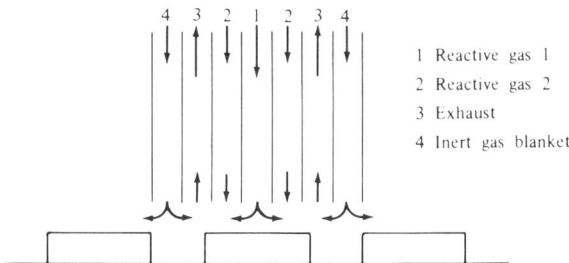

Figure 7–4: A gas-blanketed downflow system.

radio frequency (RF) energy or ultraviolet (UV) energy. In an RF-heated susceptor, the energy in an RF coil is coupled into a coated carbon susceptor. The wafers are heated through their contact with the susceptor. UV heating is accomplished by the use of high-intensity bulbs that emit strongly in the UV spectrum. The large amounts of energy from these bulbs heat the wafers and their holders by radiation. In both types of cold-wall heating, the walls of the chamber are only cold in comparison to the wafers themselves. Radiation and conduction from the susceptor produce a large temperature rise in the chamber walls. Hot-wall systems are heated using thermal resistance heating, as in a diffusion furnace. In addition to the advantages of less deposition on the walls of a cold-wall system, the wafers may be heated and cooled much more rapidly because of the small thermal mass of the system and the relatively large gas-flow velocities.

In the effluent-handling section of a CVD system, all unreacted gas plus the carrier gas must be exhausted in some manner. Generally, the exhaust gases are cleaned of any harmful or reactive gases, cooled, and vented to the atmosphere.

7.1 | CVD Methods

Boundary layer theory may be used to describe the dynamics of the basic CVD process. The theory first presumes that there is a layer of gas at the surface (boundary) of the substrate where the gas velocity is zero. The various reactants must first diffuse across this surface and then react with the atoms of the substrate at the surface. The reaction products, on the other hand, must diffuse back through this boundary layer in order to escape. The actual diffusion rates of the reactants and reaction products depend upon the following parameters in the manner shown:

1. Temperature, $T^{3/2}$
2. Total Pressure, $\frac{1}{P_T}$
3. Reactant Partial Pressure, P_n

The thickness of the boundary layer, in turn, depends on the following parameters in the manner shown:

1. Distance in the direction of gas flow, $x^{1/2}$
2. Velocity of the gas, $1/V$

Among the important considerations in operating a CVD reactor are uniform temperature profile, proper substrate positioning in gas flow, and gas flow con-

7.1 CVD Methods

Table 7–1: CVD Deposition Methods

METHOD	TEMPERATURE (°C)	MATERIALS	USES	FILM QUALITY
Atmospheric CVD	300–500	SiO_2, Phosphorus glass	Passivation, dielectric	Good
Low-temperature LPCVD	300–500	SiO_2, Phosphorus glass	Passivation, dielectric	Good
Medium-temperature LPCVD	500–900	Poly-Si, SiO_2, Phosphorus glass, Si_3N_4	Gate conductor, dielectric, passivation	Excellent
Plasma-enhanced CVD	100–300	SiO_2, SiN	Passivation, dielectric	Poor

trol. In addition, care must be taken not to introduce contaminants such as oxygen and water vapor from leaks, which will cause competing reactions.

Low-temperature CVD processes for depositing phosphorus glass and silicon nitride for passivation are especially popular because these films can be deposited directly over metal films. One drawback, however, is the resulting poor step coverage. This becomes a severe drawback when such a film is used as an insulator between two conducting layers. Higher temperature processes in the 500 to 900°C range will provide more conformal step coverage and are more suited for multilayer structures. Table 7–1 summarizes the important technologies for depositing doped and undoped dielectric and polysilicon films. Low-pressure CVD is rapidly becoming the dominant CVD technique due to its superior step coverage. In addition, low-temperature processes prevent the movement of shallow junctions, so the higher temperature processes are being replaced by LPCVD and plasma-enhanced depositions.

Almost all of LPCVD is done utilizing a horizontal quartz diffusion tube as the reaction chamber, as shown in Figure 7–5. The quartz tube is sur-

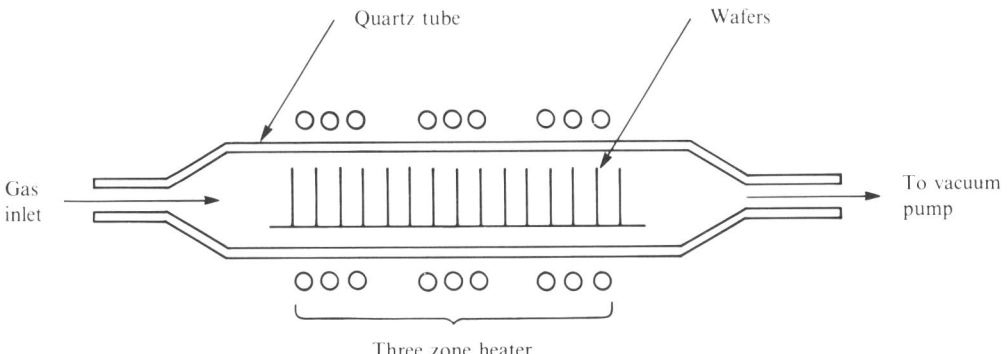

Figure 7–5: An LPCVD reactor.

rounded by a resistance heater with a mass flow controller on one end and a vacuum pump on the other. The main advantage of the horizontal reactor is its ability to process large wafer batches. An accompanying drawback, however, is the depletion of the reactant species in the gas as it flows over the wafers, causing the deposition rate to decrease. Ramping the temperature along the length of the process tube helps solve this problem but may also increase the film grain size at higher temperatures. Some reactors use a flat temperature profile and multiple gas injection ports along the tube to provide uniform deposition rates. The barrel type reactor previously described provides more uniform gas flow and lower particulate generation than the horizontal reactor, but at the expense of lower throughput.

7.2 CVD Procedures and Uses

CVD can be used to deposit many materials, but in semiconductor processing the materials generally encountered, besides epitaxial silicon, are:

1. Polycrystalline silicon
2. Silicon dioxide (both doped and undoped)
3. Silicon nitride

Each of these materials may be deposited in a variety of ways, and each has many applications.

Polycrystalline silicon is silicon with only a short-range crystal structure. It may be deposited if the deposition rate on a substrate is high, if the substrate has no crystal structure, or if the deposition temperature is below the threshold for single-crystal growth. Two general methods for deposition of polycrystalline or "poly" are:

REACTION	CARRIER GAS	DEPOSITION TEMPERATURE (°C)
$SiH_4 + Heat \longrightarrow Si + 2H_2$	H_2	850–1000
$SiH_4 + Heat \longrightarrow Si + 2H_2$	N_2	600–700

The crystal structure of the poly depends on both the deposition temperature and the rate, and may be tailored for a particular application. Polycrystalline silicon is usually deposited undoped and is doped later in the processing to provide a conductive layer for use in devices. It may be deposited doped if required. The thickness of a polycrystalline layer may be determined by interference techniques.

7.2 CVD Procedures and Uses

Silicon dioxide is obtained by using any of the following reactions:

REACTION	CARRIER GAS	DEPOSITION TEMPERATURE (°C)
$SiH_4 + 4CO_2 \longrightarrow SiO_2 + 4CO + 2H_2O$	N_2	500–900
$2H_2 + SiCl_4 + CO_2 \longrightarrow SiO_2 + 4HCl + C$	H_2	800–1000
$SiH_4 + 2O_2 \longrightarrow SiO_2 + 2H_2O$	N_2	200–500

Silicon dioxide may also be deposited so as to contain arsenic, phosphorus, or boron by respectively including some arsine, phosphine, or diborane in the reaction. These impurities form dopant oxides that are readily incorporated into the layer of deposited silicon dioxide. Deposited oxide may be used as a predeposition source if it is doped, or as a masking or implant barrier; but its primary use is as a scratch protection layer over already completed circuits and metallization. To avoid problems with the metals already deposited, the deposition is generally performed at temperatures below 450°C. A multilayer structure consisting of alternate phosphorus-doped layers and undoped oxide layers or of alternate undoped layers, phosphorus-doped layers, and undoped oxide layers is often used, as shown in Figure 7–6.

The multilayered structure is needed because:

1. The phosphorus-doped oxide is a chemical barrier to prevent the movement of contamination through the layer. It also reacts with water to produce an electrolytic etching action on the underlying aluminum metallization in the presence of an applied voltage if the percent phosphorus by weight is greater than 5%.

2. To prevent the etching action, an undoped layer of SiO_2 may be used either below, or above and below, the phosphorus-doped layer of SiO_2.

The thickness of a deposited layer is determined using the same techniques as for thermally grown SiO_2. Deposited SiO_2 is not as "dense" as thermally grown SiO_2, but heating it to 900°C or higher for 30 minutes produces properties that make the two kinds of SiO_2 almost indistinguishable.

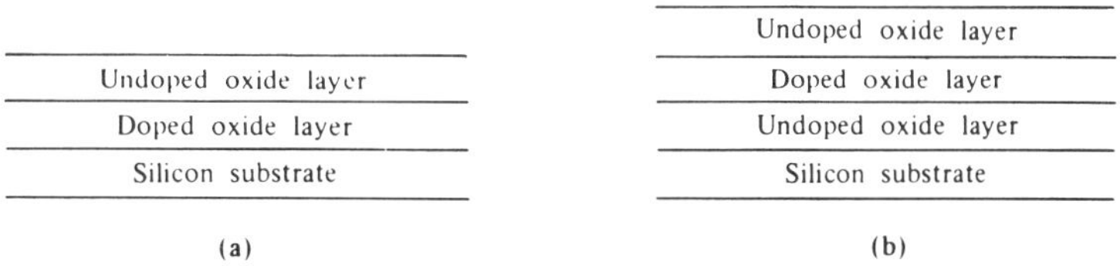

Undoped oxide layer
Doped oxide layer
Silicon substrate

(a)

Undoped oxide layer
Doped oxide layer
Undoped oxide layer
Silicon substrate

(b)

Figure 7–6: Cross section of silicon dioxide deposited for scratch protection. (a) Doped oxide capped with undoped oxide; (b) Doped oxide with undoped oxide both above and beneath.

The concentration of phosphorus in a layer of deposited SiO_2 is often determined by simultaneously depositing the layer on a lightly doped p-type silicon wafer. The wafer is then diffused for a predetermined time in a furnace set to a standard temperature. A four-point probe reading is sufficient to determine whether the deposited SiO_2 contains the desired amount of phosphorus. The actual concentration may be obtained from Figure 7-7, which is a graph of the sheet resistance of a wafer following a 30-minute predeposition at 1000°C versus the phosphorus concentration in the deposited oxide layer.

Silicon nitride is a dense dielectric that is often used to passivate circuits with device parameters that are sensitive to contamination. It may also be used for the controlled local oxidation of silicon. Silicon nitride can be deposited using CVD techniques as follows:

REACTION	CARRIER GAS	DEPOSITION TEMPERATURE (°C)
$3SiH_4 + 4NH_3 \longrightarrow Si_3N_4 + 12H_2$	H_2	900–1100
$3SiH_4 + 4NH_3 \longrightarrow Si_3N_4 + 12H_2$	N_2	600–700

The thickness of a deposited layer of silicon nitride (Si_3N_4) can be determined fairly accurately through the use of a color chart, just as SiO_2 thicknesses are determined. But, because the optical properties of Si_3N_4 differ from those of SiO_2, a different relationship between the observed color and the film thickness exists. This relationship is shown in Table 7-2.

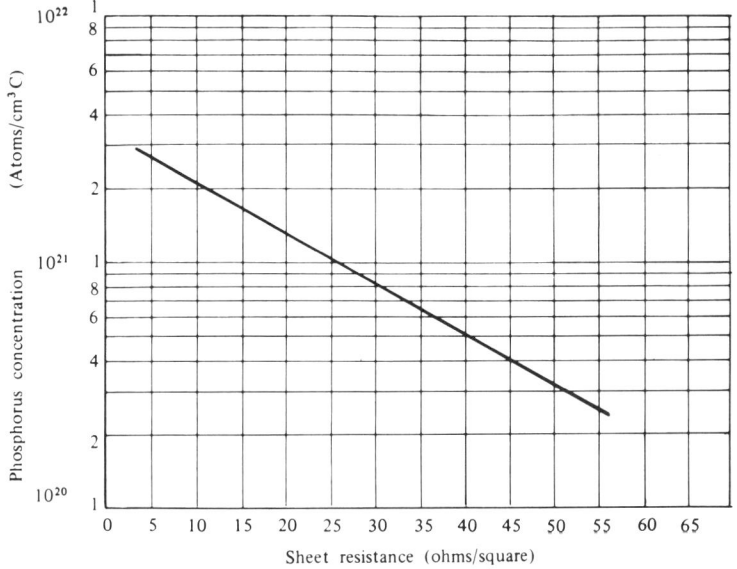

Figure 7-7: Phosphorus concentration in deposited oxide layer versus resistance following a 30-minute diffusion in nitrogen at 1000°C.

Table 7–2: Color Chart for Thermally Grown Si_3N_4 Films (Films Observed Perpendicularly Under Daylight Fluorescent Lighting)

FILM THICKNESS Å	μ	COLOR AND COMMENTS
380	.038	Tan
530	.053	Brown
750	.075	Dark violet to red violet
900	.090	Royal blue
1130	.113	Light blue to metallic blue
1280	.128	Metallic to very light yellow green
1500	.150	Light gold or yellow—slightly metallic
1650	.165	Gold with slight yellow orange
1880	.188	Orange to melon
2030	.203	Red violet
2250	.225	Blue to violet blue
2330	.233	Blue
2400	.240	Blue to blue green
2550	.255	Light green
2630	.263	Green to yellow green
2700	.270	Yellow green
2780	.278	Green yellow
2930	.293	Yellow
3070	.307	Light orange
3150	.315	Carnation pink
3300	.330	Violet red
3450	.345	Red violet
3530	.353	Violet
3600	.360	Blue violet
3680	.368	Blue
3750	.375	Blue green
3900	.390	Green (broad)
4050	.405	Yellow green
4200	.420	Yellowish (not yellow, but where yellow is expected)
4280	.428	Light orange
4350	.435	Borderline between light orange or yellow to pink
4500	.450	Carnation pink
4720	.472	Violet red

(continued)

Table 7–2: Color Chart for Thermally Grown Si_3N_4 Films (Films Observed Perpendicularly Under Daylight Fluorescent Lighting) (continued)

FILM THICKNESS Å	μ	COLOR AND COMMENTS
5100	.510	Borderline between violet and blue green; looks greyish
5400	.540	Blue green to green (quite broad)
5780	.578	Yellowish
6000	.600	Orange
8200	.820	Salmon
8500	.850	Dull light red violet
8600	.86	Violet
8700	.87	Blue violet
8900	.89	Blue
9200	.92	Blue green
9500	.95	Dull yellow green
9700	.97	Yellow to yellowish
9900	.99	Orange
10,000	1.00	Carnation pink
10,200	1.02	Violet red
10,500	1.05	Red violet
10,600	1.06	Violet

A complete summary for atmospheric, low-pressure, and low-pressure plasma-enhanced CVD is given in Table 7–3.

Table 7–3: Deposition of Semiconductor Materials

		ATMOSPHERIC-PRESSURE CVD		
FILM	REACTANT GASES (CARRIER)	TEMP °C	DEPOSITION RATE, Å/MIN	COMMENT
Epitaxy Si CW	$SiCl_4/H_2(H_2)$ $SiH_2Cl_2(H_2)$ $SiHCl_3(H_2)$ $SiH_4(H_2)$	1125–1200 1050–1100 1100–1150 1000–1075	500–2000 500–3000 500–2000 100–300	SOS Rate 2 μm/min
Poly Si CW	$SiH_4(H_2)$	850–1000	1000	14 wafers/load 40 3″/HR
Si_3N_4 CW	$SiH_4/NH_3(H_2)$ $SiH_2Cl_2/NH_3(N_2)$	900–1100	100 230	40 3″/HR Poor control

Table 7–3: Deposition of Semiconductor Materials (continued)

FILM	REACTANT GASES (CARRIER)	TEMP °C	DEPOSITION RATE, Å/MIN	COMMENT
ATMOSPHERIC-PRESSURE CVD				
SiO_2 CW	$SiH_4/O_2(N_2)$	200–500	700	PH_3 doped 160 3"/HR
LOW-PRESSURE CVD (0.5–1.0 TORR)				
Epitaxy Si	$SiH_2Cl_2(H_2)$	1000–1075	400–650	30–80 Torr
	$SiH_4(H_2)$	900–1000	150–300	30–80 Torr
Poly Si	100% SiH_4	610	100	SiH_4 Flow Sensitive 100 3"/HR
	23% SiH_4	640	190	High Deposition Rate 150 3"/HR
Si_3N_4	$SiH_2Cl_2/NH_3(N_2)$	750	40	100 3"/HR
		900	80	
	$SiH_4/NH_3(N_2)$	840	30	100 3"/HR Spacing Sensitive
SiO_2	SiH_2Cl_2/N_2O	900	120	
	SiH_4/N_2O	860	50	Low deposition rate Si rich
SiO_2	SiH_4/O_2	450	100	
	$SiH_4/PH_3/O_2$	450	120–180	Very dependent on spacing; PH_3 doped 50 3"/HR
LOW-PRESSURE PLASMA-ENHANCED CVD				
Si_3N_4	$SiH_4/NH_3(N_2)$	300	300–350	0.2 Torr

Review Exercises

1. Briefly describe the difference between a hot-wall and a cold-wall CVD system.

2. Name three nonepitaxial materials that can be deposited using CVD techniques.

3. Following a 30-minute predeposition at 1000°C, your test wafer has a sheet resistance of 35 Ω/square. Determine the phosphorus concentration in the layer of deposited SiO_2.

4. Explain the purpose of the reaction chamber in chemical vapor deposition.

5. List and describe the five main sections of a chemical vapor deposition system.
6. Explain the difference between epitaxial growth and chemical vapor deposition.
7. Name and describe a reaction that may be used to deposit silicon nitride.

8 | Metallization

8.0 | Introduction

After the devices in the silicon substrate have been fabricated, they must be connected together to perform circuit functions. This connection process is called *metallization* and is performed using one of several available vacuum deposition techniques. In this chapter, the requirements of metallization systems, methods of depositing metals and other materials, and additional considerations in metallization are covered.

8.1 | Metallization Requirements

To serve as an effective interconnect metallization for silicon, the metal chosen must meet or exceed all of the following requirements:

1. The metal must make a low-resistance electrical contact with the silicon.
2. The metal must have a limited reactivity with silicon, in order to produce a stable contact.
3. The metal must have high electrical conductivity, so that high current is easily carried without any voltage drops.
4. The metal must adhere well to the underlying silicon, silicon dioxide, or other dielectric used.
5. A pattern must be easily definable in the layer using photolithographic techniques.

6. The deposition method must be compatible with already existing structures.
7. The metallization must uniformly cover steps in the surface topography.
8. The metallization must not exhibit "electromigration." (Electromigration is the migration of atoms in the metallization caused by the flow of current.)
9. The metallization must not corrode under normal operating conditions.
10. It must be possible to bond easily to the metallization, in order to allow for external connection.
11. The metallization must be economically competitive.

No one metal perfectly meets all of these requirements. However, aluminum does meet all of them satisfactorily. Accordingly, aluminum is the metal most often chosen for device interconnection. Recent work has shown that the performance of aluminum is improved by the introduction of small amounts of other elements. The tendency of aluminum to react with silicon is halted by introducing a small percentage of silicon in the aluminum during deposition. In a similar fashion, the electromigration resistance of aluminum is greatly increased by including a small percentage of copper in the aluminum layer during deposition. These elements are added as follows:

1. *Copper.* The addition of 3 to 5 percent copper to aluminum increases the electromigration resistance of the aluminum by a factor of 10 to 100.
2. *Silicon.* The addition of 1 to 2 percent silicon in aluminum prevents the aluminum from dissolving the silicon in the substrate, which can cause "spiking" when shallow junctions are present.

In those cases where aluminum does not meet the requirements needed for the metallization layer, multilayered metallization schemes are often utilized. Each layer will meet some of the requirements previously covered, and the combination of the layers satisfies all of them.

8.2 | Vacuum Deposition

A metal layer on a wafer is most often obtained using a vacuum deposition technique. There are many types of vacuum deposition systems, but they all have some characteristics in common. To perform any type of vacuum deposition, a system must have the following:

1. A chamber that may be evacuated to provide a sufficient vacuum for the deposition to take place. (The chamber must include valves, etc., for the job.)

8.2 Vacuum Deposition

2. A vacuum pump (or pumps) to reduce the gas pressure in the chamber to an acceptable level.
3. Instrumentation to monitor the vacuum level and other system parameters.
4. A method of depositing the desired layer or layers of material.

Each of these needs may be met in many ways, but many trade-offs are involved. A typical vacuum deposition system is shown in Figure 8–1.

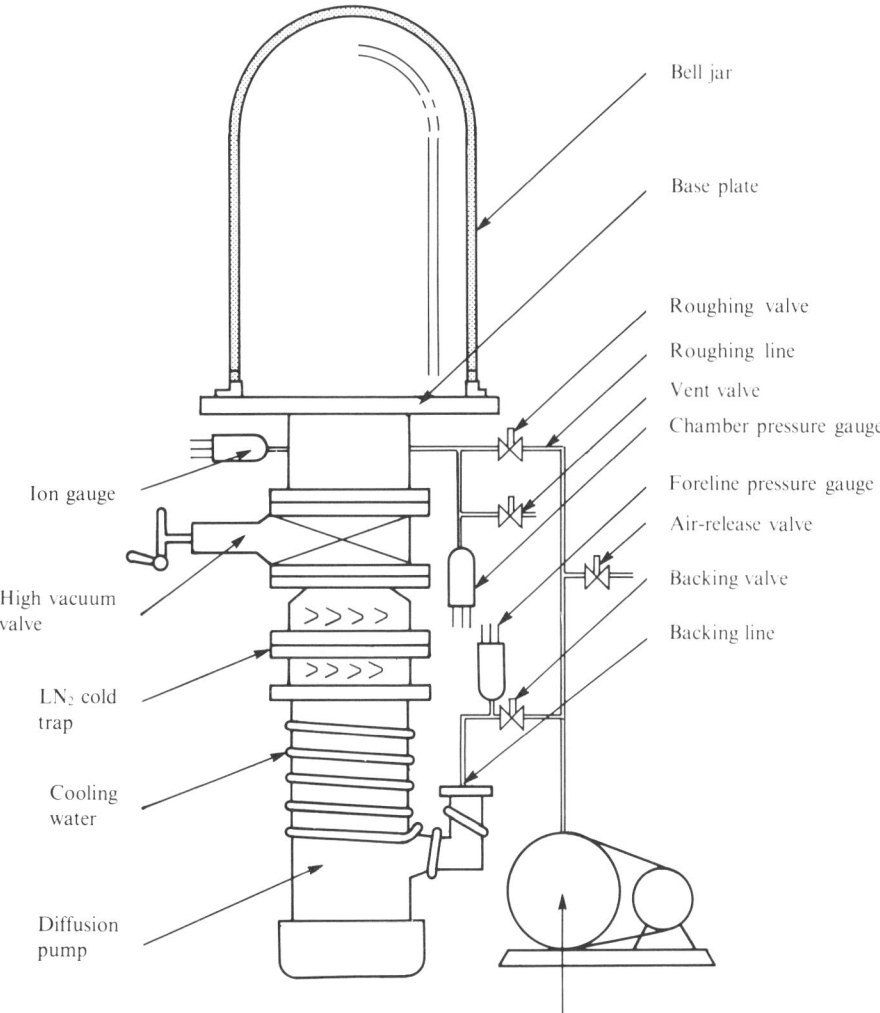

Figure 8–1: Schematic diagram of a typical fast-cycling, high-throughput vacuum coating system.

A vacuum chamber consists of a leak-free enclosure that allows sufficient access for the deposition fixturing and instrumentation. Both glass and stainless steel enclosures are common, but stainless steel accommodates more nonstandard configurations and does not break.

To obtain a sufficient vacuum, different types of pumps are used. Pumps are like gears on a car—different types work more efficiently over different vacuum ranges. The following is a summary of different pump types and the pressure ranges over which they are used.

1. Atmospheric pressure to intermediate vacuum levels (10–100μ).
    a. *Rotary oil-sealed pumps*. The rotary oil-sealed pump uses a rotor that is sealed against leakage by a vacuum oil. The air left in the vacuum system enters the pump through the inlet port, is compressed, and is ejected to the atmosphere through the exhaust or discharge port. A schematic of a vane-type rotary oil pump is shown in Figure 8–2.
    b. *Sorption pump*. The sorption pump uses chemicals that will adsorb gases on their surface. Containers of these chemicals adsorb gases until no more can be accommodated (the process usually takes many cycles), at which point they must be baked out to restore their capacity to adsorb. This type of pump is illustrated in Figure 8–3.
2. Intermediate vacuum levels to low vacuum levels (25μ–10^{-6} mm).
    a. *Diffusion pump*. In the diffusion pump, vapor from a boiler passes through a series of nozzles in a downward direction, carrying resid-

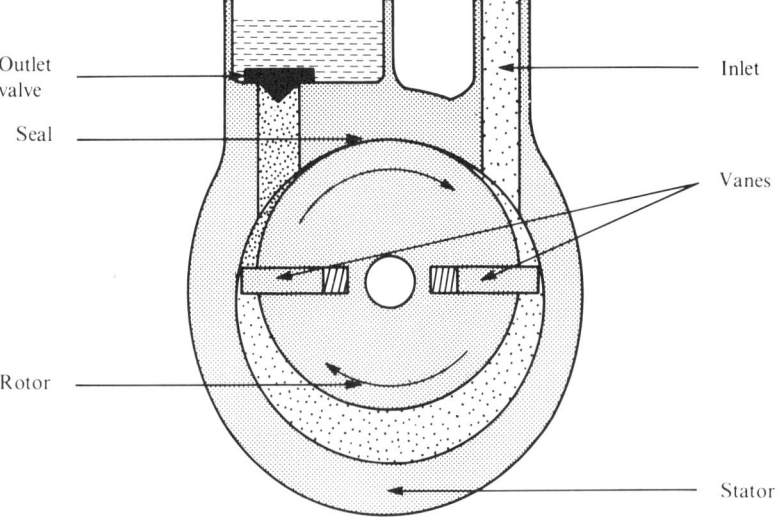

Figure 8–2: Vane-type rotary oil pump. *(After Maissel and Glang, Ref. 1.)*

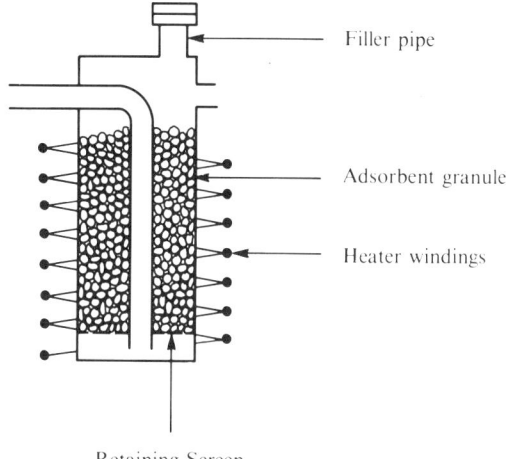

Figure 8–3: Bakeable foreline sorption pump. *(Ref. 2.)*

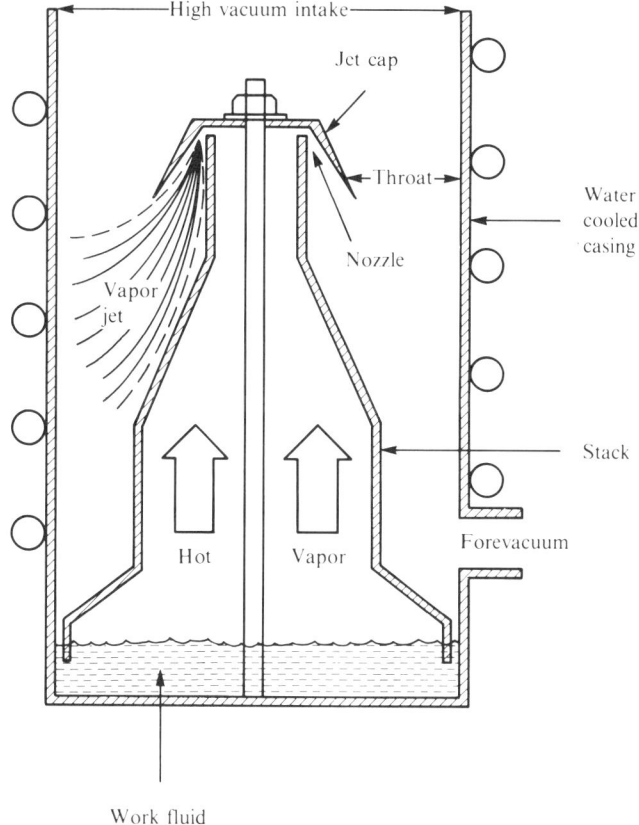

Figure 8–4: Cross section of an oil diffusion pump. *(Ref. 3.)*

139

ual atoms in the vacuum chambers with it. The basic elements of a diffusion pump are shown in Figure 8–4.
   b. *Turbomolecular pump.* The turbomolecular pump has a series of blades set many levels deep around a hub (in similar fashion to an electrical turbine) in order to propel molecules out of the chamber by imparting suitable momentum to them. The pump is illustrated in Figure 8–5.
   c. *Cryogenic pump.* The operating components of the cryogenic pump are cooled metal surfaces in the form of disks, tubes, or cylinders. These surfaces act to trap those residual gas molecules that strike them. Figure 8–6 is an illustration of a cryogenic pump.
3. Low vacuum levels to ultra low vacuum levels ($10^{-6}10^{-10}$ mm).
   a. *Ion pump.* Using a combination of an electric and a magnetic field, the ion pump provides a method of ionizing atoms and then trapping them. Figure 8–7 illustrates such a pump.

The instrumentation of a vacuum deposition system must provide a method of:

1. Determining the vacuum level in the chamber.
2. Measuring the status of all valves, etc., in the system.
3. Determining the thickness of any deposited layers.

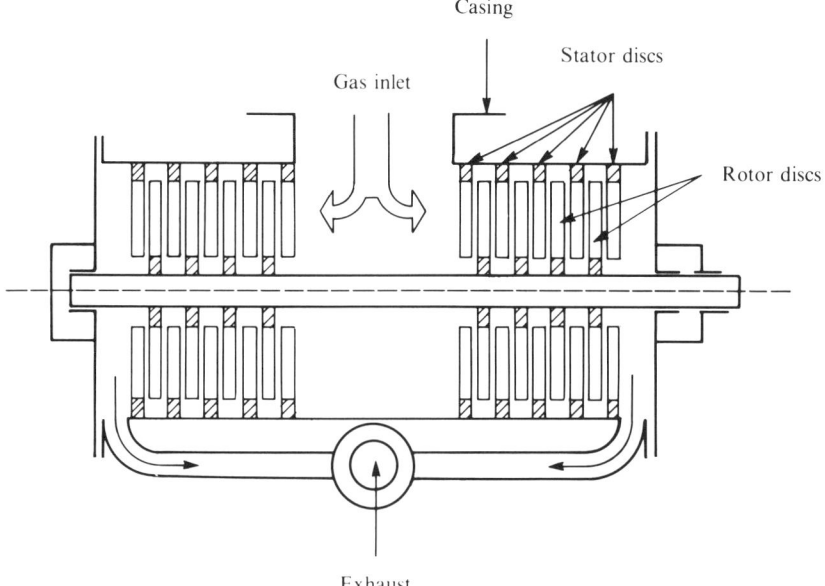

Figure 8–5: Cross section of a turbomolecular pump. *(Ref. 4.)*

8.2 Vacuum Deposition

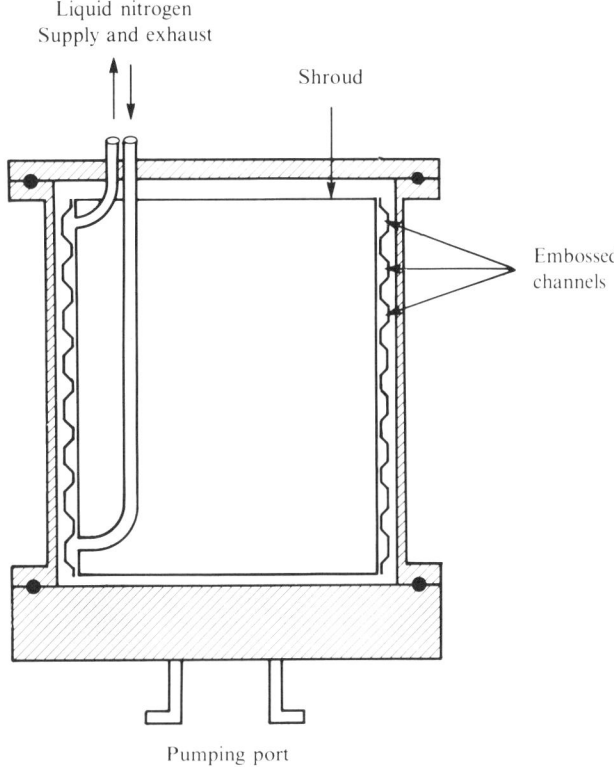

Figure 8–6: Cross section of a cryogenic pump. *(Ref. 5.)*

The vacuum level inside the vacuum chamber may be determined from atmospheric pressure to intermediate vacuum levels using a diaphragm that moves with changes in pressure. A mechanical or electrical readout is used. For better vacuums, a measure of the ability of the residual gas to carry heat away from a filament is often used. The more gas present, the more heat can be carried away. This principle is used in both the Thermocouple and the Pirani Gauge. A third technique for lowering vacuum levels uses an ionization tube. Atoms entering the tube are ionized by a heated filament, and then an electric field accelerates these ions toward an electrode where they are collected. The resulting current flow is proportional to the vacuum level.

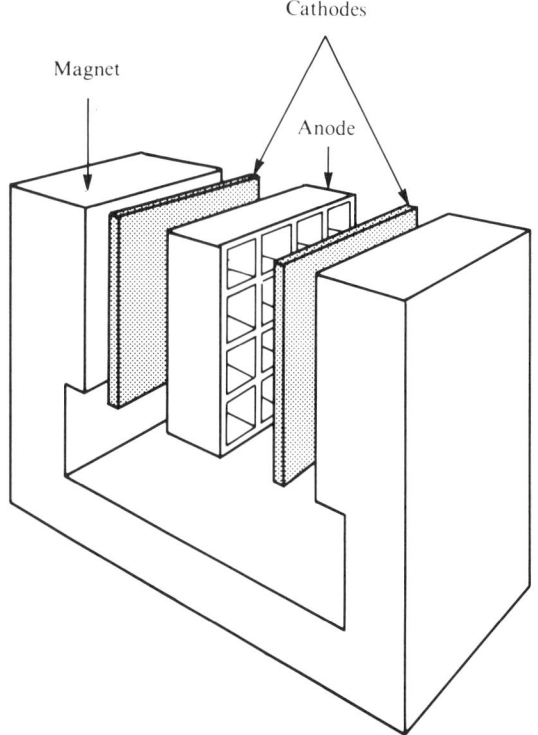

Figure 8–7: Basic elements of a sputter-ion pump. *(Ref. 6.)*

8.3 | Deposition Techniques

Four vacuum deposition techniques are used to deposit materials in the semiconductor industry:

1. *Filament evaporation.* Current flows through a filament causing first melting and then evaporation of the material on the filament.
2. *Electron-beam (E-beam) evaporation.* An intense beam of electrons is used to evaporate the material to be deposited.
3. *Flash evaporation.* A wire or pellets are fed onto a hot ceramic substrate, evaporating the material on contact.
4. *Sputtering.* A gas at low pressure, such as argon, bombards and dislodges atoms from a target of the material to be deposited.

Each of these methods has advantages and disadvantages, and the trade-offs involved must be considered when selecting among them.

8.3 Deposition Techniques

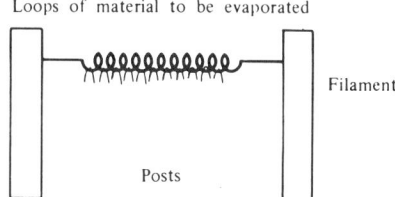

Figure 8–8: A typical filament evaporation system.

Filament evaporation is the simplest and least expensive deposition method. Evaporation takes place from a filament or a boat heated by thermal resistance heating. Figure 8–8 shows a typical filament evaporation source.

Evaporation is accomplished by gradually increasing the current flowing through the filament to first melt the loops of material, thereby wetting the filament. (Care must be taken to choose a filament compatible with the material to be evaporated.) Once the filament is wetted, the current through it is increased to accomplish the evaporation. Filament evaporation systems are easily set up, and many materials can be evaporated using them. However, the contamination level of the deposited materials is often sufficiently high to interfere with the functioning of the device. Contamination may come from the filament or from poor handling techniques. For this reason, filament evaporation of aluminum is not common. Care can be taken to minimize this effect, but other methods have proved more economical and reliable. Filament evaporation is often used to deposit backside gold, however, since contamination is not of concern in this case. The technique cannot be used to evaporate composite materials because the element with the lowest melting point evaporates first, leaving the rest of the material to evaporate later.

Electron-beam (frequently called E-beam) evaporation uses a focused beam of electrons to heat the material. A high-intensity beam of electrons is generated in a manner similar to that used in a television picture tube. The focused beam of electrons melts the material contained in a hearth, a water-cooled block with a large depression for holding the evaporation source. Only electrons come in contact with the material to be evaporated, so it can be a low-contamination process. E-beam evaporation is also a rapid process, but it cannot be used for the deposition of composite materials unless more than one hearth is employed. Because of the intense electron-beam source used, radiation damage may occur to the devices in their substrates. This damage must be annealed out later in the process. A typical electron-beam evaporation source is shown in Figure 8–9.

Flash evaporation is similar to filament evaporation in that the material is evaporated by thermal resistance heating, but the similarity ends there. Flash evaporation uses a continuously fed spool of wire (or in some cases, a stream of pellets or powder) incident on the heated ceramic bar for the deposition, as

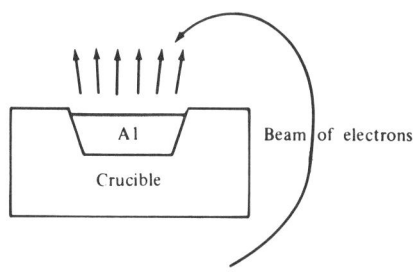

Figure 8–9: A typical electron-beam evaporation system.

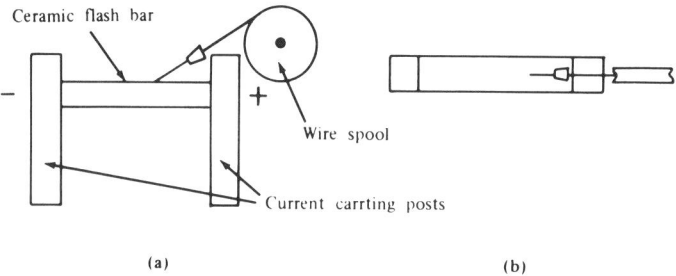

Figure 8–10: A flash evaporation system (a) Side view; (b) Top view.

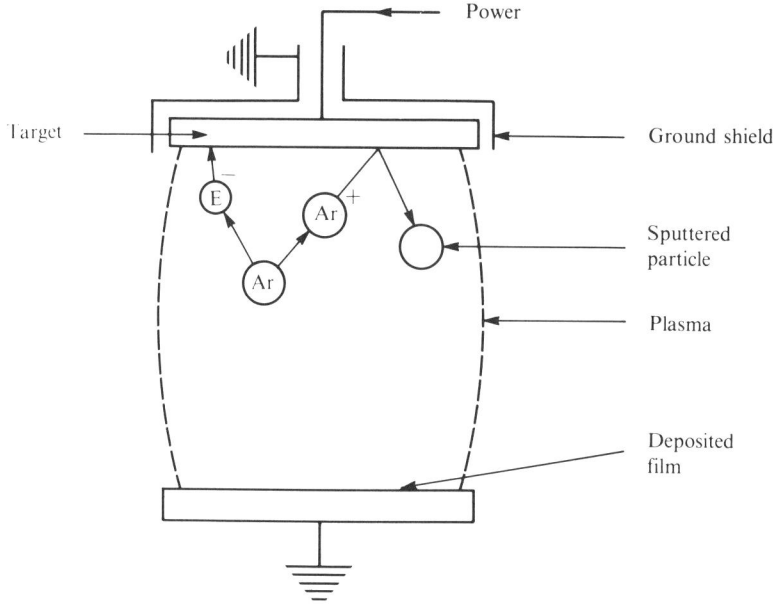

Figure 8–11: A sputtering system.

shown in Figure 8–10. This deposition technique combines the speed and contamination-free features of E-beam deposition with the radiation-free feature of filament evaporation and still offers the option of depositing composite layers.

In the sputtering method of vacuum deposition, ions of an inert gas such as argon are introduced into the chamber after a satisfactory vacuum level has been reached. An electric field ionizes these atoms and accelerates them toward one electrode in the chamber, called the target. When the ions strike the target, they dislodge atoms from it, depositing them on the substrates facing the target, as shown in Figure 8–11. Sputtering can be accomplished using both DC and RF voltages, and it can be used to deposit almost any material, although the deposition rate is often extremely low. The addition of a magnetic field at the target electrode, a technique called *magnetron sputtering*, increases the deposition rate of metals considerably.

Table 8–1 summarizes the various methods commonly utilized in the semiconductor industry to deposit different metal films.

8.4 Vacuum Deposition Cycle

The following metallization sequence is typical, regardless of the particular material being deposited or type of equipment being used in a given manufacturing facility:

1. Clean the wafers to remove all contamination, etc., and dry them.
2. Position the wafers in the vacuum chamber so that they will receive a uniform layering. (In many cases, a rotating structure called a *planetary* is used. The planetary rotates and revolves in the chamber, guaranteeing that the coverage of steps on the surface is as good as possible.)
3. Close the vacuum chamber and "rough" down to ~25μ.
4. Close the valve to the roughing pump, and open the valve to the high-vacuum pump. Allow the system to reach the required vacuum level (10^{-6} to 10^{-7} mm is typical).
5. Turn on the source and evaporate a small amount of material onto a shield between the source and the wafers to clean the source.
6. Deposit the necessary thickness of material on the substrates. (The substrates may be heated to increase adhesion and step coverage.)
7. Turn the source off and cool the wafers and the chamber.
8. Fill the chamber with an inert gas like N_2, open it, and remove the wafers.

Table 8–1: Metal Deposition Characteristics

	DEPOSITION RATE	FILM THICKNESS	COMPOSITE MATERIALS	CONTAMINATION	STEP COVERAGE	DEGRADES DEVICE?	TYPICAL APPLICATIONS	COST RANGE
Filament	Medium	Thin	No	Yes	Good	No	Back metal	$40K–100K
E-beam	High	Thick	No (unless multiple source)	No	Good	Yes (X-rays)	Aluminum interconnect	$60K–140K
Flash	High	Thick	Yes	No	Good	No	Al–Cu–S: interconnect	$50K–120K
Sputter	Low (magnetron sputtering high for metals)	Thick	Yes	No	Excellent	Slight (high-energy ions)	Al–Cu–S: interconnect	$100K–300K

8.5 Evaluation of Film Characteristics

The two parameters of interest in evaluating a deposited metal film are thickness and resistivity. The most popular method for measuring sheet resistance is the four-point probe method. In this method, four aligned osmium probes are brought into electrical contact with the conductive film. A known current I is then applied through the outside probes, creating a current flow and resultant electrostatic fields between the probes. The resulting drop in voltage V is measured between the inner probes. The calculation of V/I produces a value of measured resistance.

One method of measuring thickness is to etch the conductor film to produce a sharp step, and then measure the thickness of the film at the step using contact surface profiling or interferometric techniques. An alternative method involves placing the conductive metal film between two RF coils and measuring the changes in the quality factor Q of the circuit. The sheet resistance R_S is measured directly, as with the four-point probe. The thickness of the conductive film is given by

$$t = \frac{\rho}{R_S} \tag{8-1}$$

where ρ is the resistivity of the film in Ω-cm and R_S is the sheet resistance of the film in Ω/square. To calculate the resistivity of a conductive film, the thickness and sheet resistance must be determined independently. Sheet resistance is measured by the RF method, and thickness is again measured by preparing an etched step in the conductive film and then using a surface profiler.

Review Exercises

1. List five requirements that a metallization must meet.
2. List and briefly describe four vacuum deposition techniques.
3. Which vacuum deposition technique may lead to radiation damage?
4. What is the name of the rotating structure that wafers are mounted on during vacuum deposition?
5. Why is aluminum the most frequently used metal for the process of metallization?

6. Why are trace amounts of silicon and copper added to the aluminum during metallization?
7. Name the major components of a vacuum deposition system.
8. List and explain four methods of depositing metals using vacuum techniques.
9. Describe a typical vacuum deposition system.

REFERENCES (CHAPTER 8)

1. Maissel and Glang, *Handbook of Thin-Film Technology*, McGraw-Hill, 1970, pp. 2–5.
2. Ibid., pp. 2–7.
3. Ibid., pp. 2–10.
4. Ibid., pp. 2–8.
5. Ibid., pp. 2–19
6. Ibid., pp. 2–35.

9 | Device processing: from alloy to sale

9.0 | Introduction

The remaining processing steps that devices undergo between metallization and final sale are as important as the initial steps. However, the "back end" of the line does not have the glamour of the rest of the processing sequence. But, as the price of silicon chips continues to fall, the companies that package, test, and distribute the devices most efficiently will remain strong in the marketplace. The device flow for the remainder of a processing line is discussed below.

9.1 | Alloying–Annealing

The successful etching of the aluminum on the front side of the wafer to form the device interconnection does not guarantee that a good electrical contact has been formed. A subsequent *alloying* step is usually used to produce a low-resistance contact between the aluminum and the silicon. Alloying is performed in a diffusion furnace set at a relatively low temperature. The alloy temperature and time vary from process to process, but the limits on temperature can be determined in part by looking at the diagram of the aluminum–silicon system in Figure 9–1.

In the figure, the line indicated by the arrows shows the lowest temperature at which a completely molten mixture exists. This temperature varies with the atomic percent of silicon in the aluminum, as indicated by the figure. The lowest temperature at which a molten solution exists is the one at the intersection of the two lines, at 577°C. This temperature is the aluminum–silicon *eu-*

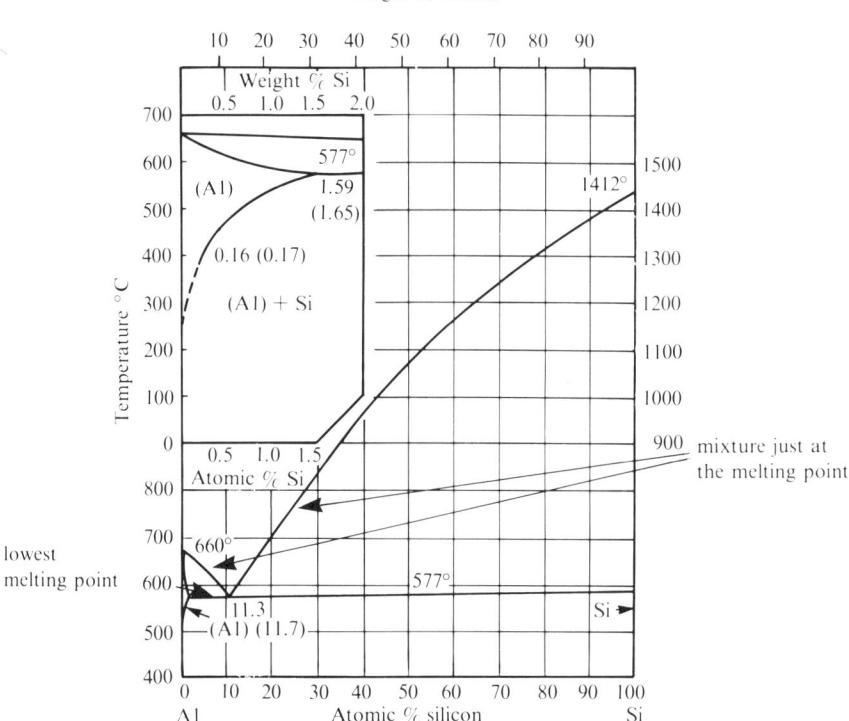

Figure 9–1: The aluminum–silicon system.

tectic temperature. If an aluminum–silicon mixture is heated to above 577°C, melting will occur, ruining any devices that are present. The upper limit in the alloy temperature is thus 577°C. The lower temperature is set by process considerations like cleanliness and the aluminum deposition temperature. Most alloying steps are performed at temperatures between 450 and 550°C for times of between 10 and 30 minutes.

During alloying or directly following it, the wafers are often heated in a gas mixture containing hydrogen (or occasionally another gas). This step is usually called *annealing*. Annealing is designed to optimize and stabilize device characteristics. Hydrogen is thought to combine with uncommitted atoms at or near the silicon–silicon dioxide interface, thus reducing their effect on device performance. Typical annealing temperatures are 400 to 500°C for times of 30 minutes to 60 minutes.

The resistance R of the resulting aluminum–silicon contact depends upon:

1. Contact area
2. Surface doping concentration

9.1 Alloying—Annealing

3. Surface cleanliness
4. Alloy or sintering conditions

The goal of the aluminum alloy or sintering process is to minimize the contact resistance. If alloying is performed at the correct temperature, and if the surface is properly cleaned, the dependence of the contact resistance can be reduced to:

1. Contact opening size
2. Doping concentration

Figures 9–2 and 9–3 indicate the dependence of contact resistance R_c in Ω-cm^2 upon doping concentration for aluminum on p-type and n-type silicon,

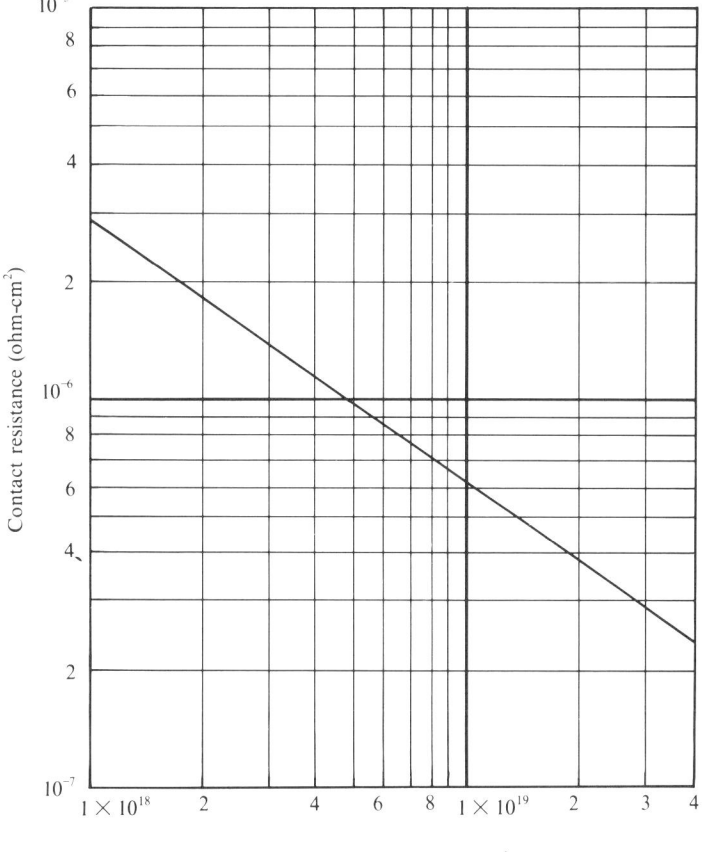

Figure 9–2: Contact resistance of aluminum on *p*-type silicon.

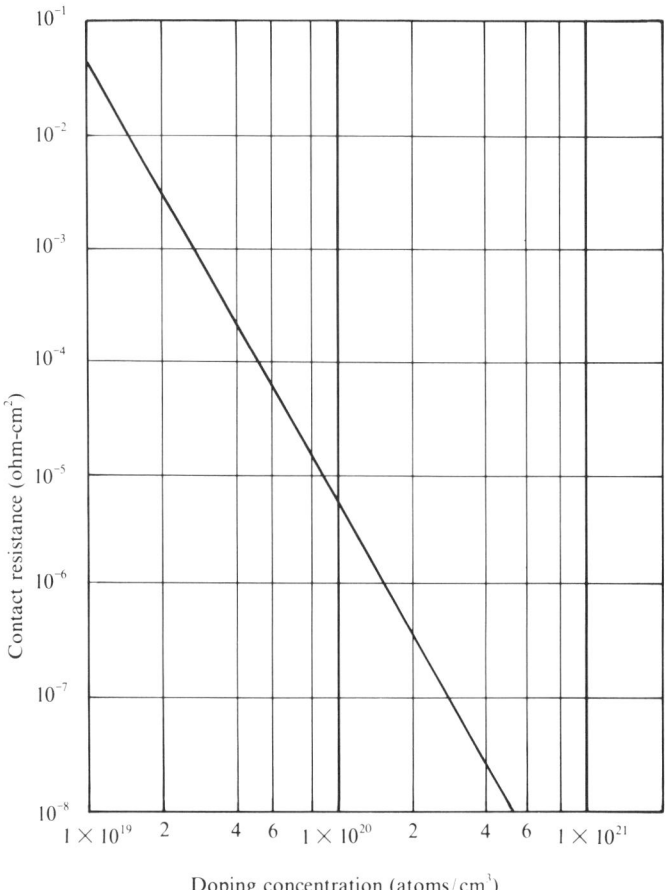

Figure 9–3: Contact resistance of aluminum on *n*-type silicon.

respectively. Once the contact dimensions are known, it is a simple matter to calculate the actual resistance R from equation 9–1 for a rectangular contact:

$$R = \frac{R_c}{(\text{contact length}) \times (\text{contact width})} \tag{9-1}$$

9.2 Post-Alloy Sample Probing

Following metallization, etching, alloying, and annealing, the wafers should contain fully functional devices. If the wafers contain only discrete devices, such as diodes or transistors, it is easy to test a certain fraction of the devices

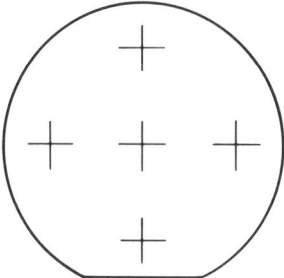

Figure 9–4: Sample areas on a typical wafer.

to see if they function properly. If more complicated integrated circuits have been fabricated, it is usually necessary to test some of the diodes, resistors, transistors, etc., that must all work for the circuit to function properly. In this manner, it is possible to discard wafers that have no chance of containing an economical number of good dies. However, this probing step provides more than just a quick check on the wafers from the fabrication line. By properly selecting the devices that are tested, and by taking data beyond a pass or fail level, it is possible to trace variations that are occurring in the fabrication process. For instance, variations in the value of a base resistor may indicate an unwanted change in the amount of boron in the base of a transistor. This change may lead, in turn, to a change in the gain of the transistors that are being fabricated. By anticipating variations in transistor gain, it may be possible to correct for a small problem before it becomes a major one.

Post-alloy sample probing (sometimes called "electrical testing") is often performed using a piece of test equipment called a curve tracer together with a probe station. On a typical wafer, the performance of devices from different areas of the wafer will be measured as shown in Figure 9–4. Measurements taken using this or a similar pattern provide information on processing variations that exist across the wafer.

9.3　Scratch Protection

To protect the devices on the wafers from improper handling and chemical contamination, a layer of CVD silicon dioxide or silicon nitride is generally used. The deposition considerations for this layer were discussed in Chapter 7. Following deposition, openings in the deposited layer are etched over the bonding pad areas. The wafers are now ready for any backside preparation that is necessary prior to wafer sorting.

9.4 | Back-Side Preparation

The back side of a wafer may have to be altered to prepare for subsequent processing steps. Two back-side preparation steps that are common are:

1. *Back-side lapping.* The back side of a wafer is lapped to remove diffused layers that interfere with its electrical properties, to thin the wafer, to make it easier to separate the die, or to prepare the back side for subsequent metal deposition.

2. *Back-side metal deposition.* A metal such as gold is deposited on the wafer back to make the attachment of the separated die to the package easier in a later operation.

Neither of these steps, one of these steps, or both of these steps may be performed on wafers to prepare them for later operations.

When a metal is used for back-side contact, it is usually deposited using filament evaporation. Gold is often chosen for the back-side metal because of the low gold–silicon eutectic temperature, shown to be 370°C in Figure 9–5. This temperature is sufficiently low to prevent degradation of other device

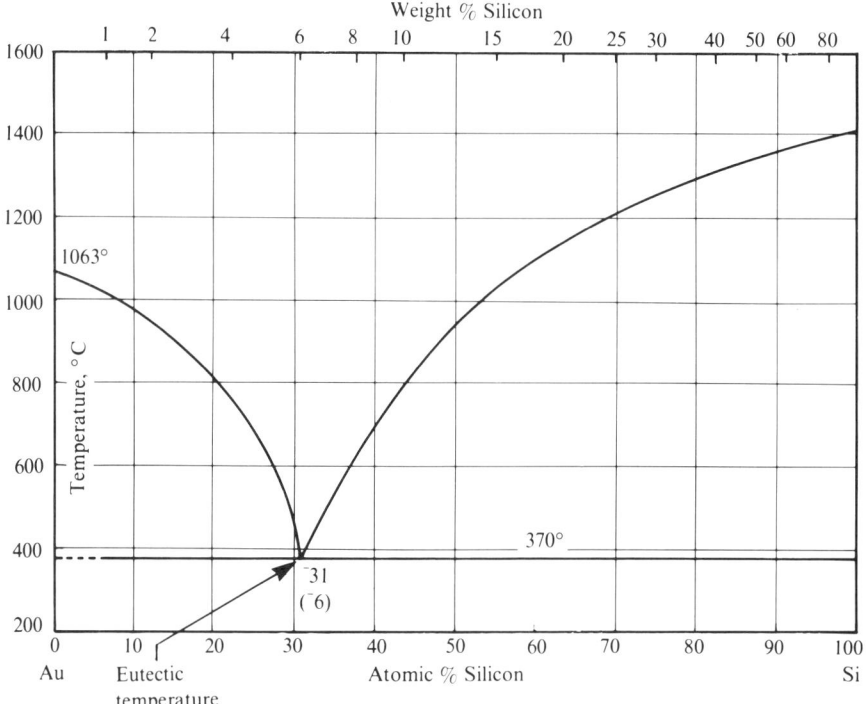

Figure 9–5: The gold–silicon system phase diagram.

characteristics when the gold is alloyed to the silicon and when the die is attached to a package.

9.5 | Wafer Sort

The wafers are now at their moment of truth—do they contain any functional die? To determine the matter, the wafers are placed on a (usually computer-controlled) wafer prober, and the individual die on each wafer are tested. Pointed metal probes contact each bonding pad and supply the necessary currents and voltages to the device. Those devices that function properly are left alone, while those that fail are marked with a drop of ink from an "inker." In some instances, die may be differentiated as top-grade devices and ordinary-grade devices. In this case, two different types of ink may be employed to differentiate between the grades. The ink is usually baked to prevent unwanted removal at a later step.

9.6 | Device Separation

With the devices or circuits on the wafer tested, it is time to separate the wafers into individual die for final packaging. This operation is generally called *wafer scribe*, although changes in the methods used to accomplish the separation have made this term technologically obsolete. The three methods commonly used to separate die are as follows:

1. *Diamond scribing*. A tool with a precisely shaped diamond imbedded in the tip is drawn across the wafer along the scribe line, making a "mark" or "scribe" in the wafer. The imperfection in the crystal structure caused by the scribing defines the crystal planes along which the wafers tend to break. By bending the wafer on both sides of the scribe line, the wafer is broken along the line.

2. *Laser scribing*. A laser is pulsed to generate a series of holes in the silicon wafer along the scribe line, as shown in Figure 9-6. The series of holes serves as the line along which the wafer breaks. This technique is relatively recent and is occasionally complicated by the condensation of the silicon initially evaporated by the laser. (The term often used is "kerf.") Back-side laser scribing and the use of a protective layer of material are two ways of preventing kerf from affecting the device yield.

3. *Sawing*. An even more recent development than laser scribing is the use of a rotating blade to separate the die. Recent advances in metallization have made it possible to manufacture saw blades capable of separating die with a minimum of silicon loss. Use of this method provides die that have uniform dimensions and square sides. These features make the process attractive for automated die-handling techniques.

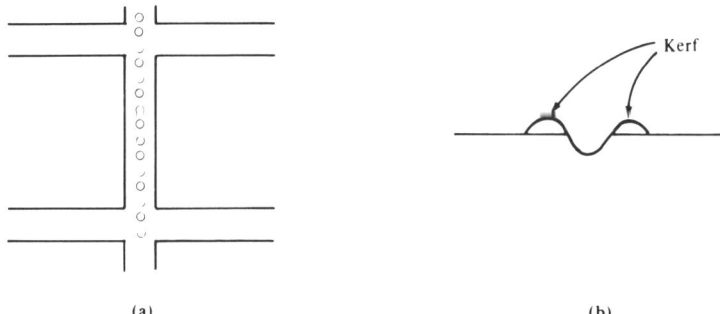

Figure 9–6: The pattern left by a laser scriber. (a) Top view; (b) Side view.

9.7 | Die-Attach (Die Bonding)

For convenience of handling, the good die are attached to a package or a leadframe. Four types of die-attach are encountered: eutectic, preform, soft-solder, and epoxy. *Eutectic* die-attach involves the prior deposition of a layer of metal, such as gold, on the backside of the die. By heating the package to above the eutectic temperature (370°C for gold-silicon) and placing the die on it, a bond is formed between the die and the package. *Preform* die-attach involves the use of a small piece of material of special composition that will adhere to both the die and the package. A preform is placed on the die-attach area of a package and allowed to melt. The die is then "scrubbed" across the region until it is attached, and then the package is cooled. Both eutectic and preform die-attach are similar to welding in that part of the die melts and solidifies to form the bond. *Soft-solder* die-attach uses a metal, such as silver, deposited on the wafer back to form the bond. A lead–tin preform melts at the die-attach temperature, bonding the die to the package. Soft-solder die-attach is similar to soldering in that the metal melts, but the silicon in the wafer does not. *Epoxy* die-attach involves the use of an epoxy glue to attach the die to the package. A drop of epoxy is dispensed on the package, and the die is placed on top of it. The package may need to be baked at an elevated temperature to cure the epoxy properly. Epoxy die-attach can be performed using either electronically conductive or nonconductive material, while the other two die-attach methods form conductive bonds only.

9.8 | Wire Bonding

The type of wire used to electronically connect the bonding pads of the device to the package is usually aluminum or gold. Gold is considerably more expensive than aluminum but offers the advantages of corrosion resistance and higher current-carrying capability. Three methods of attaching the wires between the device pads and the packages are *thermo-compression* (TC) bonding, *ultrasonic* (US) bonding, and *thermosonic* (TS) bonding. In each case, the name of the bonding method is an accurate description of the steps involved. Thermo-compression bonding is used with gold wire and involves heating the package and forming the bond between the wire and the pad using both heat and pressure. Ultrasonic bonding uses a pulse of ultrasonic energy to provide a scrubbing action that forms a bond between the aluminum wire used and the pad. Thermosonic bonding uses gold wire and a combination of heat and a pulse of ultrasonic energy.

9.9 Packaging Considerations

Semiconductor processing technology has evolved to the point that the cost of the package may be a considerable fraction of the total cost of the device. For this reason, packaging considerations are receiving increasing attention. The prime consideration is the material used to construct the package. The oldest and generally most reliable package is made of either metal or metal and ceramic. These packages also tend to be the most expensive, so replacements for them are constantly being sought. The use of various plastic and epoxy packages has become popular in recent years because of the low cost and the ease of forming the package. The results obtained with these packages are better every year, but they still cannot match the metal or metal-and-ceramic package in terms of protecting the die from the environment.

Another consideration is the ability of a package to conduct heat away from the die. Special metal tabs, fins, or wings may be designed as part of the package to conduct heat away from the die while it is in operation.

9.10 Final Test

Once a device is packaged, it is ready for final testing. The test the device undergoes may well be the one it underwent after wafer sort, but the handling steps involved in bonding and packaging the device may have damaged the die, or these steps may not have been performed correctly. If either has occurred, the packaged device will not perform properly; hence, this test is necessary.

9.11 Mark and Pack

Once nonfunctional devices have been removed, the last step prior to storing the fully functional devices, called *mark and pack*, is performed. The packages are marked with the device code and date code that tells customers when they were manufactured. The devices are now ready to sell and ship to waiting customers.

Review Exercises

1. What two compositions of gold–silicon are just at the melting point at 800°C?
2. What is the eutectic composition, in both atomic percent and percent by weight, of aluminum–silicon?
3. Is the gold–silicon or the aluminum–silicon eutectic temperature higher?
4. What are two methods of separating die?
5. State three methods of connecting wire leads from a device to a package.
6. What is the purpose of the post-alloy sample probe step?
7. Explain the two types of back-side wafer preparation.
8. Why is gold frequently used as the back-side metal?
9. How is the nonfunctionality of a die determined, and how are nonfunctional dies identified?
10. List the various steps between alloying and shipment of a completed semiconductor device.
11. A p-type resistor with 4 micrometer $\times$ 6 micrometer contact windows has a doping concentration of 2×10^{18} atoms/cm^3.
    a. Determine the R_c of the resistor.
    b. Calculate the resistance the contacts add to the value of a 100-Ω resistor.
12. An npn transistor has a base contact with dimensions of 4 micrometers $\times$ 7 micrometers and a doping concentration of 6×10^{18} atoms/cm^3. What resistance, in ohms, is added to the base resistance of the transistor due to the base contact?

10 | Device and IC technologies

10.0 | Introduction

The processing steps discussed in previous chapters are used in specific sequences to manufacture both discrete devices and integrated circuits (ICs). Many types of discrete devices and ICs are manufactured, but this chapter focuses on the two major technologies: bipolar technology and metal oxide semiconductor (MOS) technology. Although both of these technologies are based on the same processing steps, the different processing sequences and surface geometries used in each produce transistors that function on different physical principles.

10.1 | Bipolar Technology

The word *bipolar* refers to the flow of both holes and conduction electrons in the functioning of the transistor. A typical bipolar IC sequence consists of seven or more masking steps. The sequence of masks used as the basis of a bipolar IC process is:

MASK NO.	TYPE	DESCRIPTION
1	Buried layer	The heavily doped $n+$ region beneath the majority of all active devices.
1	Epitaxy	The n-type layer in which all devices are fabricated.
2	Isolation	The p-type diffused region that provides electrical isolation between adjacent regions.

(continued)

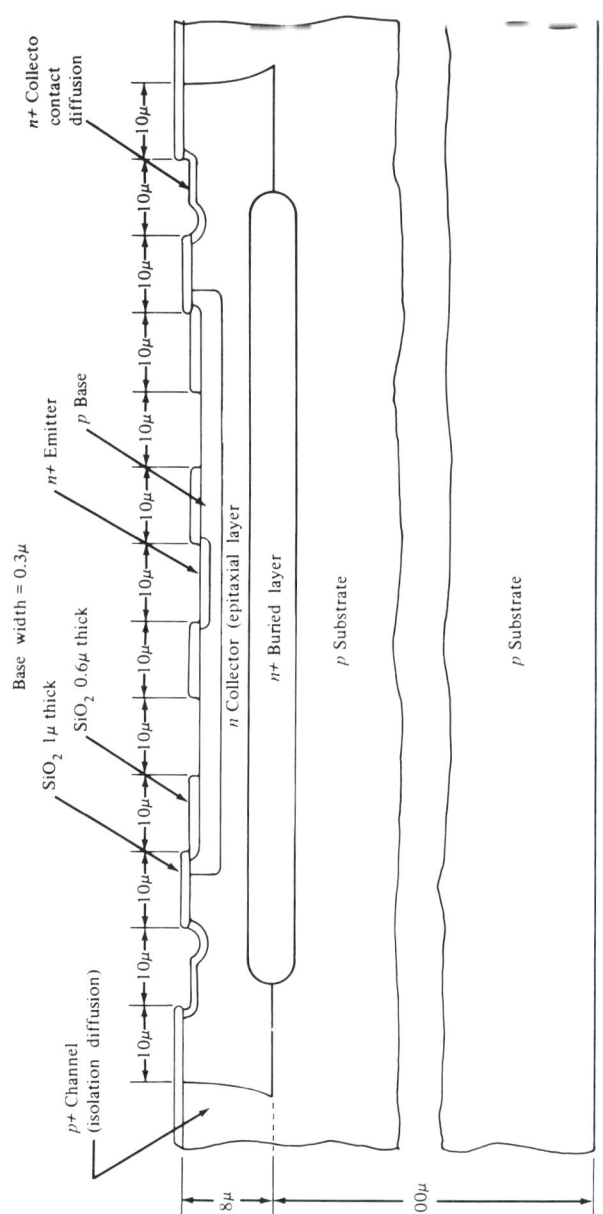

Figure 10-1: Cross section of a typical bipolar integrated circuit.

MASK NO.	TYPE	DESCRIPTION
3	Base	The *p*-type diffusion that serves as the base of all *npn* transistors and the body of most resistors.
4	Emitter	The *n*+ diffusion that forms the emitters of *npn* transistors and the contacts to their collectors.
5	Contact	Openings to provide electrical access to all devices.
6	Metallization	The conductive paths that electrically connect the devices to form a circuit.
7	Passivation	The deposited layer of SiO_2 or other material that serves as both a physical and a chemical protective barrier over the completed circuit.

A cross section of a typical bipolar circuit is shown in Figure 10–1.

10.2 Devices Fabricated Using Standard Bipolar Technology

NPN TRANSISTORS

The bipolar process sequence is optimized to fabricate bipolar *npn* transistors. These transistors are used as both amplifiers and switches in circuit designs. The current gain (also called h_{fe} or β) is the ratio of the current that flows in the collector to the current that flows in the base. The symbol for an *npn* transistor is shown in Figure 10–2; Figure 10–3 shows top and side views of a typical *npn* transistor. The current the transistor can handle is determined by the size of the device. The typical minimum-geometry transistor can handle 1–10 mA. The base–collector reverse breakdown is determined by the doping on both sides of the junction depth. Current gain is typically 50–500.

PNP TRANSISTORS

The symbol for a *pnp* transistor is shown in Figure 10–4. There are two kinds of *pnp* transistors:

1. *Lateral pnp transistor.* The current gain of a lateral *pnp* transistor is typically less than that of a vertical *npn* transistor, going to unity (i.e., no gain) at a much lower frequency than the current gain of the latter. Figure 10–5 shows top and side views of a lateral *pnp* transistor.

2. *Vertical pnp transistor.* A vertical *pnp* transistor is used if the collector of the device in the circuit goes to the circuit ground (the substrate). Figure 10–6 shows top and side views of a vertical *pnp* transistor.

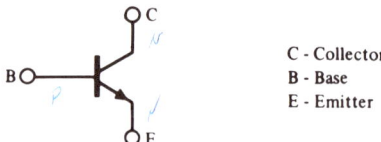

Figure 10–2: Symbol for an *npn* transistor.

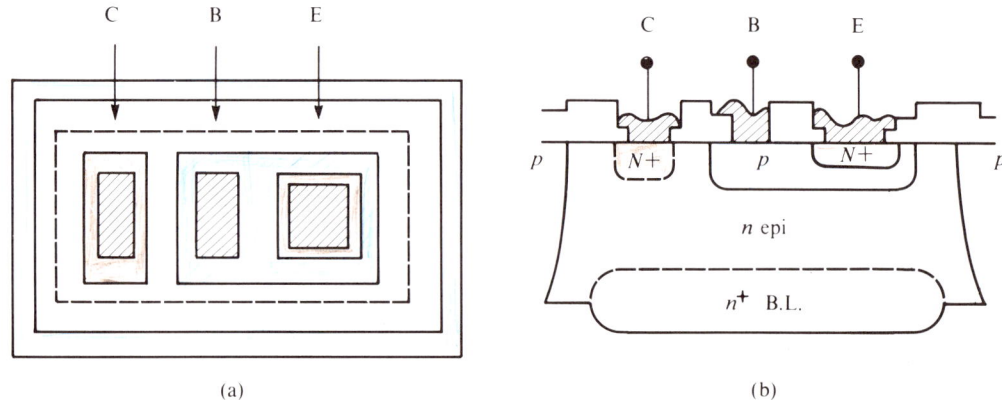

(a) (b)

Figure 10–3: An *npn* transistor. (a) Top view; (b) Side view.

DIODES

A diode is present at every *pn* junction, but only a few of the diodes are used in circuit applications. A diode allows current to flow in the direction of the arrow, but does not allow any current to flow in the reverse direction until the breakdown voltage is reached. Figure 10–7 shows the symbols for both standard diodes and Zener diodes, and the following list describes several kinds of diodes that occur in circuits.

1. *Emitter–base diode*. This device is usually formed by shorting the collector and base of an *npn* transistor together as the anode, with the emitter as the cathode. The emitter–base diode has a low (6–10 V) reverse breakdown and is often used as a Zener diode. The device is shown in Figure 10–8.

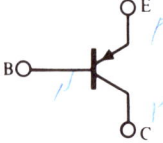

Figure 10–4: Symbol for a *pnp* transistor.

10.2 Devices Fabricated Using Standard Bipolar Technology

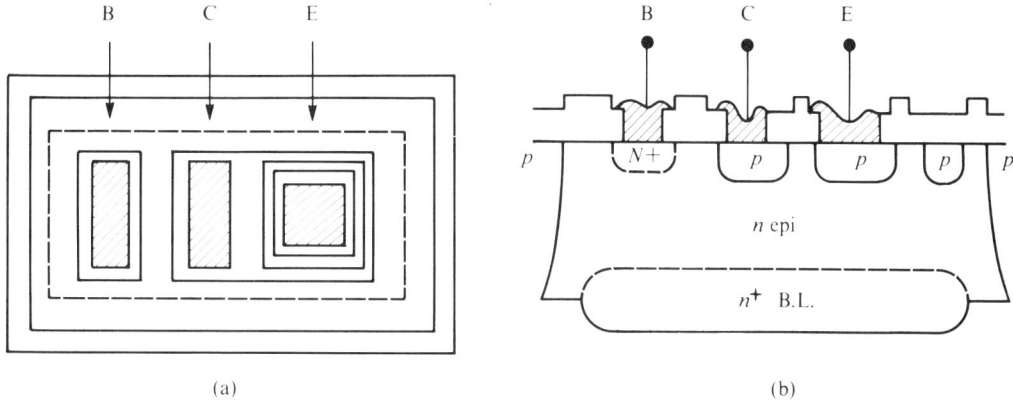

Figure 10–5: Lateral *pnp* transistor. (a) Top view; (b) Side view.

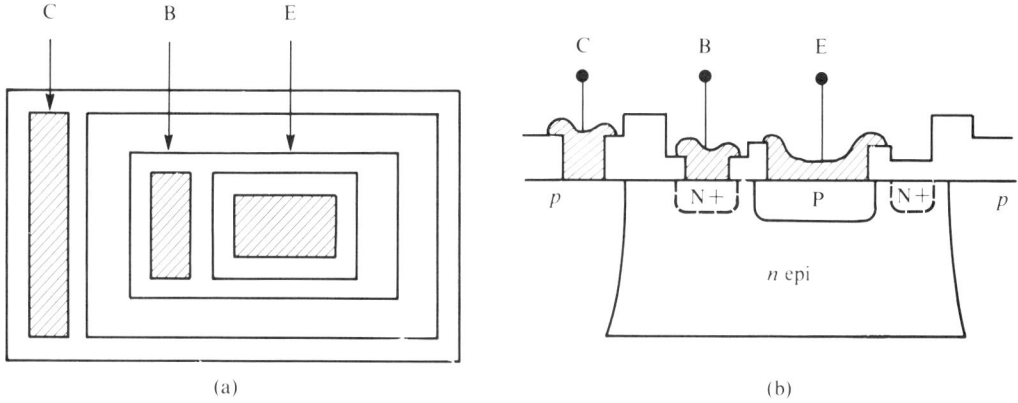

Figure 10–6: Vertical *pnp* transistor. (a) Top view; (b) Side view.

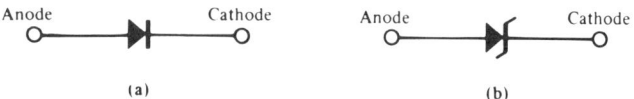

Figure 10–7: Diode symbols. (a) Standard diode; (b) Zener diode.

2. *Base–collector diode.* This diode is usually formed using the base of an *npn* transistor to form the anode, and the collector to form the cathode. Typical reverse breakdown of the device is 15–50 V. Figure 10–9 shows a typical base–collector diode.

3. *Epi-isolation diode.* These back-to-back diodes formed between any two "isolated" pockets in an integrated circuit prevent any electrical interac-

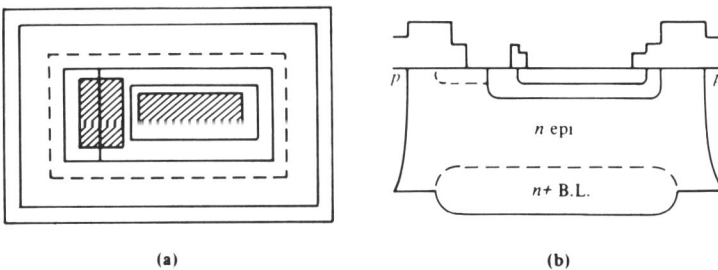

Figure 10–8: Emitter base diode. (a) Top view; (b) Side view.

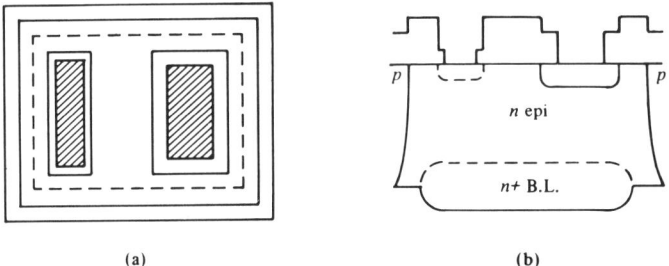

Figure 10–9: Base collector diode. (a) Top view; (b) Side view.

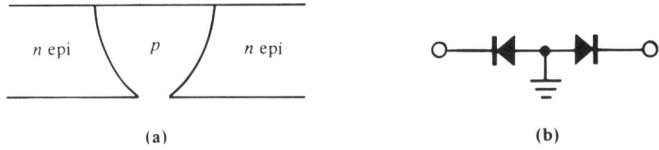

Figure 10–10: Epi-isolation diodes. (a) Side view; (b) Circuit symbol.

tion between devices in the circuit. Figure 10–10 shows the side view of epi-isolation diodes and the corresponding circuit symbol.

4. *Other diodes.* Two diodes not frequently encountered in circuit applications are the emitter–isolation diode and the isolation–buried-layer diode.

RESISTORS

Resistor symbols are shown in Figure 10–11. The current through a resistor is related to the voltage across it by the relationship

$$V = RI$$

where R is the resistance of the resistor in ohms.

10.2 Devices Fabricated Using Standard Bipolar Technology

The following list describes the different kinds of resistors used in circuits:

1. *Base resistor.* This resistor is made by contacting both ends of a region of p-type base diffusion. Typical resistance range is 50–50,000 Ω, since base sheet resistivities are generally in the range of 100–500 $\Omega/\square$. A typical base resistor is shown in Figure 10–12.

2. *Pinched-base resistor.* This resistor is fabricated by diffusing an emitter region over the center of a base resistor. Typical sheet resistance values are 2000–10,000 $\Omega/\square$, but the control on the resistance is quite poor. Typical resistor values range from 10,000–500,000 Ω. Figure 10–13 shows a typical pinched-base resistor.

3. *Emitter resistor.* A diffused emitter region is contacted at both ends to fabricate this resistor. To minimize area while preventing unwanted par-

Figure 10–11: Resistor symbols. (a) Standard symbol; (b) Pinched resistor.

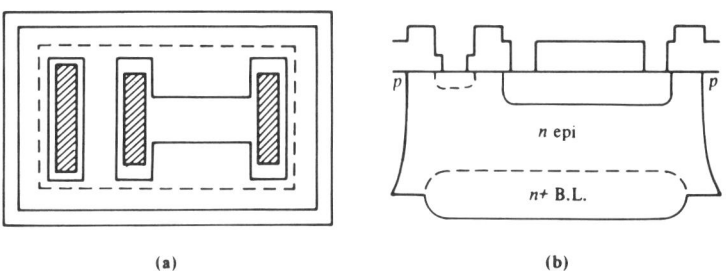

Figure 10–12: Base resistor. (a) Top view; (b) Side view.

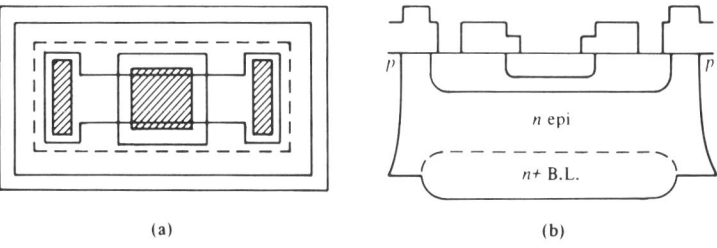

Figure 10–13: Pinched-base resistor. (a) Top view; (b) Side view.

asitic device action and retaining resistor control, the emitter diffusion is generally performed in a region that has previously received a base diffusion, and one end of the resistor is connected to the base region. Typical emitter sheet resistivities are 4–10 $\Omega/\square$, and emitter resistors of values 5–100 Ω are often made. A typical emitter resistor is shown in Figure 10–14.

4. *Epi resistor.* This resistor consists of an epitaxial region of silicon surrounded by an isolation wall with $n+$ contacts at each end. Typical sheet resistivities for epi vary from 400–2000 $\Omega/\square$. Control of epi resistor values is not as good as control of diffused resistor values because of both the variation in epi thickness and resistivity and variations in the lateral diffusion of the isolation. Typical epi resistor values are 2000–50,000 Ω. Figure 10–15 contains top and side views of an epi resistor.

5. *Pinched-epi resistor.* This resistor is similar to an epi resistor, but it has an additional p-type base diffusion over its center, reducing its current-carrying region and, hence, increasing its sheet resistance. Because of the latter property, pinched-epi resistors are often used instead of epi resistors in a design layout. A typical pinched-epi resistor is shown in Figure 10–16. Sheet resistivities range from 500–3000 $\Omega/\square$. The pinched-epi structure is used for resistors from 5000–100,000 Ω.

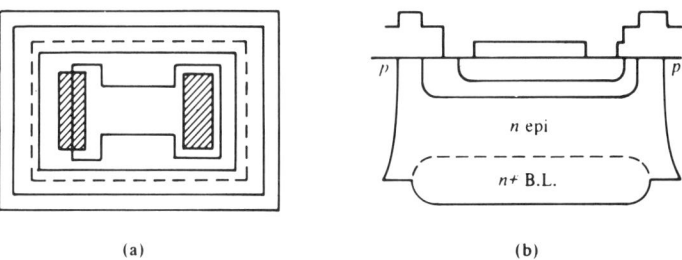

Figure 10–14: Emitter resistor. (a) Top view; (b) Side view.

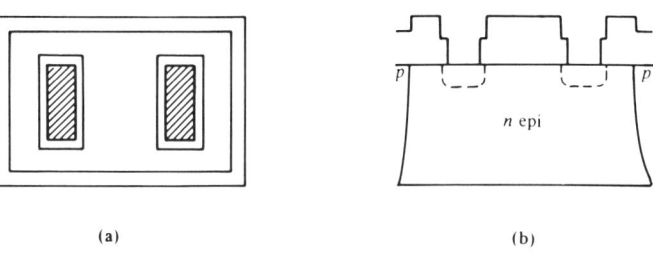

Figure 10–15: Epi resistor. (a) Top view; (b) Side view.

10.2 Devices Fabricated Using Standard Bipolar Technology

CAPACITORS

Capacitors are used in circuits for the purpose of storing charge or suppressing circuit transients. Figure 10–17 depicts the circuit symbol of a capacitor, and the following list describes two different kinds of capacitors.

1. *Dielectric capacitors.* A capacitor is formed whenever a dielectric layer separates two conductive regions. The top plate of the capacitor is most often the interconnect metallization, with one of the diffused regions (isolation, base, or emitter) used as the other plate of the capacitor. The layer of thermal oxide is the dielectric. Capacitance per unit area increases as oxide thickness decreases, so the thin emitter oxide is often used for capacitors. Figure 10–18 shows top and side views of a dielectric capacitor, with the emitter diffusion as one plate of the capacitor.

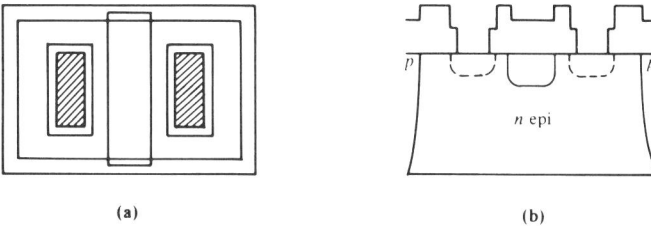

Figure 10–16: Pinched-epi resistor (a) Top view; (b) Side view.

Figure 10–17: Capacitor symbol.

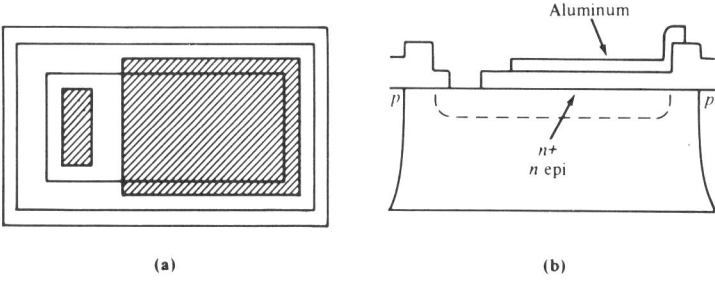

Figure 10–18: Dielectric capacitor. (a) Top view; (b) Side view.

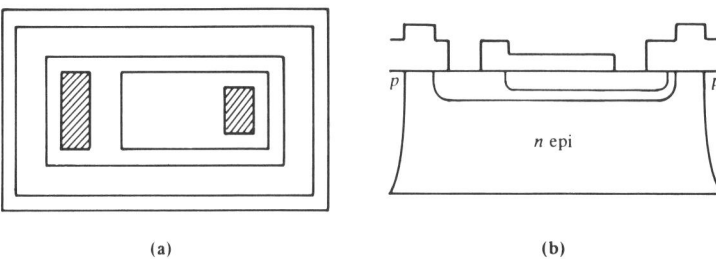

Figure 10–19: Junction capacitor. (a) Top view; (b) Side view.

2. *Junction capacitors.* A reverse-biased *pn*-junction behaves like a capacitor for small voltage excursions around the operating point. This type of capacitor is often used where the low leakage and constant capacitance of a dielectric capacitor are not needed. Figure 10–19 shows top and side views of a junction capacitor.

10.3 | Basic MOS Technology

Many variations of MOS technology exist, but the basic considerations are the same regardless of the particular steps used to fabricate the devices. Some of these steps are:

1. *Source drain.* A *p*-type or *n*-type diffusion that forms the resistors and the two current-carrying terminals of the transistors.

2. *Gate oxidation.* The careful growing of the thin SiO_2 layer that the controlling charge acts through.

3. *Contact.* The formation of openings to provide electrical access to the devices.

4. *Metallization.* The formation of the conductive paths that electrically connect the devices to form a circuit.

5. *Passivation.* Deposition of the layer of SiO_2 that serves as both a physical and a chemical protective layer over the completed circuit.

A cross section of a typical MOS circuit is shown in Figure 10–20.

The devices that can be fabricated using MOS technology include:

1. *MOS transistors.* The process is optimized to fabricate either *p*-channel or *n*-channel MOS transistors (Figure 10–21). The MOS transistor works as an amplifier or as an on–off switch in circuits.

2. *Source–drain resistor.* This is the only resistor available using the MOS process (see Figure 10–22). Typical ranges are 50–10,000 Ω, since source–drain sheet resistances are 50–200 $\Omega/\square$.

10.3 Basic MOS Technology

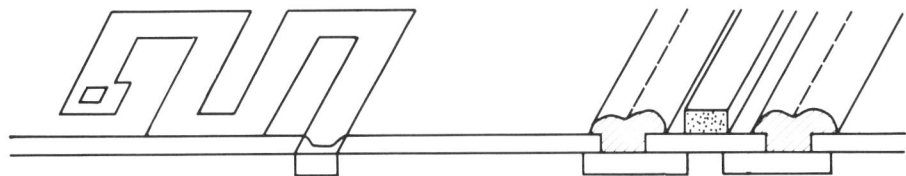

Figure 10-20: Cross section of a typical MOS circuit.

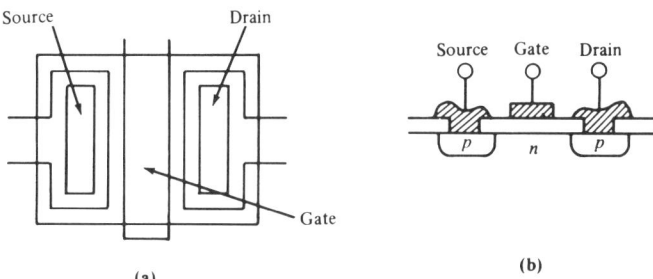

Figure 10-21: A *p*-channel MOS transistor. (a) Top view; (b) Side view.

Figure 10-22: Source/drain resistor. (a) Top view; (b) Side view.

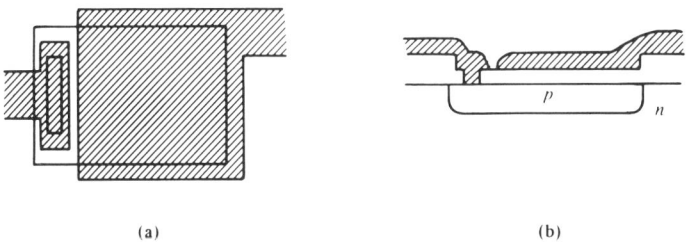

Figure 10-23: MOS dielectric capacitor. (a) Top view; (b) Side view.

3. *Capacitor.* A capacitor is formed wherever a layer of SiO_2 covers a conductive region of silicon. The capacitance per unit area increases with thinner oxide layers, so the gate oxide layer is often used for capacitors. A MOS dielectric capacitor is shown in Figure 10-23.

10.4 MOS Technology Variations

Several variations of the basic *p*-channel MOS process shown in Figure 10–21 are:

1. *n-channel MOS.* Starting with a substrate doped with the opposite conductivity type (*p*-type), *n*-type impurities are diffused into the material. This technology offers faster devices than *p*-channel MOS technology. It is called *NMOS technology.*

2. *Silicon gate.* A conductive layer of polycrystalline silicon is substituted for the metal used as the gate. This layer is also used as interconnect elsewhere in the circuit. Silicon gate techology produces devices that are both smaller and faster than metal gate technology. It is often referred to as *Si-gate technology.*

3. *SOS.* An epitaxial layer of silicon is deposited on an insulating sapphire substrate. The devices are fabricated in this thin epitaxial silicon layer. The letters stand for *s*ilicon *o*n *s*apphire.

4. *CMOS.* CMOS technology combines both *p*-channel and *n*-channel devices on the same chip by adding processing steps. Circuits are fabricated that use very little power to operate. *CMOS* stands for *c*omplementary *MOS.* The technology may become the dominant technology because of its versatility in designing new circuits and its low power dissipation. A cross section of a typical *n*-well silicon gate CMOS circuit is shown in Figure 10–24, which also shows an emerging trend—the use of a lightly doped epitaxial layer over a heavily doped substrate for CMOS circuits.

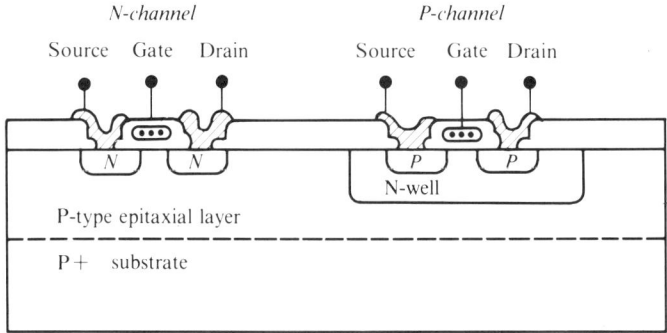

Figure 10–24: Cross section of an *n*-well silicon-gate CMOS structure fabricated in a *p/p*$^+$ epitaxial substrate. (Conventional CMOS processing does not use an epitaxial layer.)

Table 10–1: Comparison of Integrated Circuit Processing Technologies, Features, Performance, and Devices

	COMPARISON	BIPOLAR	TECHNOLOGY NMOS OR PMOS	CMOS
Process Complexity	Epitaxial Depositions Diffusion/oxidation cycles Masks in sequence	1 5 7	0 2 5	0/1 4 7
Circuit Features and Performance	Current Voltage Switching speed Power dissipation Device density	High High High High Low	Low Low Low Intermediate High	Low Low Low Low High
Devices Available for Circuit Design	Transistors Diodes Resistors Capacitors	NPN, PNP(2) At least 5 At least 5 Junction, dielectric	NMOS or PMOS 1 1 Junction, dielectric	PMOS and NMOS 3 2/3 Junction, dielectric

10.5 Comparison of Bipolar and MOS IC Technologies

In the evolution of semiconductor processing technology, bipolar ICs were introduced to the marketplace before MOS circuits became available. The strides made in processing technology in the mid 1960s led to the first commercial MOS ICs by the end of that decade. These first PMOS ICs were fabricated by means of a process sequence that was less complex than that used for bipolar ICs, as seen in Table 10-1. The less complex fabrication sequence limited the types of devices available for circuit design and produced circuits that were slower than bipolar ICs. These MOS circuits also operated over a lower voltage range and supplied less current than bipolar ICs. However, MOS circuit characteristics were ideal for digital circuit applications, such as logic devices and memories. The less complex, and hence less expensive, process sequence used for MOS circuits led to an explosive growth in the number of ICs of this kind that were produced and sold every year. By the mid 1970s, NMOS technology began displacing PMOS technology because of the inherent speed advantage of n-channel transistors. The decreasing feature size of NMOS technology in the late 1970s led to a fundamental limitation—excess chip temperature due to high-power dissipation in normal circuit operation. CMOS technology, with its very low power dissipation (see Table 10-1) began to be used in new designs.

The advantages in speed, current, and voltage that bipolar technology had when MOS technology was first developed have diminished over the years. As both bipolar and MOS technology have evolved, the simple structures and process sequences shown in this chapter have become considerably more complex. Bipolar technology still offers the best approach to solving many circuit problems, but MOS technology, particularly CMOS technology, is being used in many areas previously reserved for bipolar technology. The coming years will see continuing development in both of these technologies, but as they evolve, circuits composed of both bipolar and MOS devices may be used to solve circuit problems not addressed by either process technology alone.

Review Exercises

1. What are the two categories that fabrication technologies are divided into?

2. How do the number of masking steps in bipolar and MOS technology compare?

3. How is a diode formed from a bipolar *npn* transistor?
4. Does a base resistor or a pinched-base resistor have a higher resistance per unit area?
5. Draw the cross-sectional view of a *p*-channel MOS transistor.
6. What is the purpose of the buried layer of the *npn* transistor in the bipolar process?
7. What is the purpose of the gate oxide in the MOS transistor structure?
8. What are the advantages of dielectric capacitors over junction capacitors?
9. What devices are fabricated using MOS technology?
10. What devices are fabricated using bipolar technology?

11 | The wafer fabrication environment

11.0 | Introduction

Throughout the entire semiconductor fabrication process, it is critical to minimize the amount of contamination that comes in contact with the wafers and the wafer processing equipment. Indeed, contamination control has become a major factor in manufacturing yield and profitability in semiconductor processing. Contamination from the facility itself, and from factory personnel, the processing equipment, and all the materials used to fabricate the devices, are major concerns in the manufacturing environment.

A theoretical model which is often used to estimate the impact of contamination on yield is given by

$$Y = \frac{1}{(1 + AD)^N} \tag{11-1}$$

where

Y = the yield
A = the device chip area in square inches
D = the defect density in defects per square inch
N = the number of masking steps.

This equation is plotted in Figure 11–1 for values of D.

A couple of conclusions may be drawn from the figure. First, probe yield decreases dramatically as die size increases. Second, as the number of defects on the wafer (defect density) increases, the yield exhibits a corresponding de-

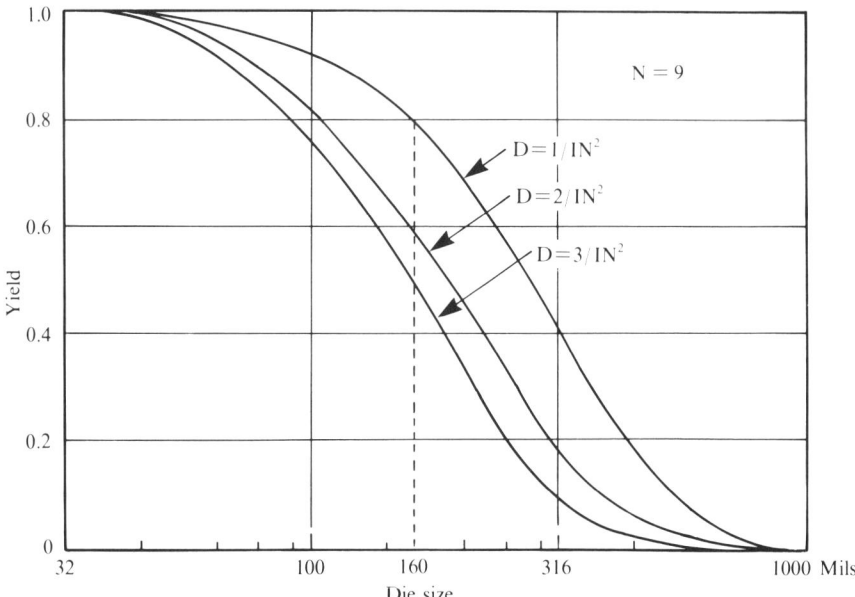

Figure 11–1: Die yield for a nine-mask-level process.

crease. Thus, adequate monitors and controls are necessary in order to fabricate semiconductor devices with reasonable yields.

Semiconductor processing materials may introduce a wide range of contaminants during the process sequence. Table 11–1 lists the major processing materials and major sources of contamination for each.

Table 11–1: Materials and Contaminant Sources

PROCESSING MATERIAL	CONTAMINANT SOURCE
High-Purity Water	Ionized species Bacteriological growth Organic residues Colloidal matter
Acids and Solvents	Chemical integrity Inherent particles Transferred particles
Photoresists	Gel particles Packaged particles
Bulk Gases	Transferred particles Chemical contaminants
Specialty Gases	Cylinder contaminants Leak integrity, valves Chemical variations

11.1 Chemicals and Cleaning Procedures

One of the first decisions to be made in wafer processing is the cleaning method to be used prior to any process step. Attempts are made to keep wafers clean at all times, but prior to high-temperature processing steps such as diffusion, epitaxial growth, or chemical vapor deposition, even more care must be taken. The two types of contamination found to cause the largest problems in semiconductors are ions that are mobile in silicon dioxide, e.g., sodium, and elements that diffuse in silicon and precipitate out in the wafer interior, e.g., gold and some other metals.

Sodium interferes with the normal operation of semiconductor devices by rapidly drifting through silicon dioxide towards a region with negative voltage. It then gives rise to changes in device characteristics, such as excessive leakage. Sodium may be kept out of a fabrication line by specifying low-sodium chemicals and by rigorously enforcing proper wafer-handling techniques. Sodium is a chemical present in the human body, and careless procedures will result in unwanted sodium contamination of wafers or wafer-handling equipment.

Certain elements are soluble in silicon at elevated temperatures, but precipitate into nonlattice locations when the wafer temperature is lowered. These elements interfere with the normal flow of holes and electrons in the silicon crystal when the device is in operation. Once a quantity of any of these elements has contaminated a wafer, it is impossible to completely remove them. However, proper cleaning prior to a high-temperature operation minimizes the probability of introducing such contaminants. Numerous methods of cleaning are popular in the semiconductor industry, but they all have certain common characteristics. The first step in dealing with wafers with an unknown history is to thoroughly degrease them. A common method is the use of a degreasing chemical such as 1,1,1-trichloroethane followed by rinses in acetone and alcohol. This cleaning procedure guarantees that any greases or waxes that might be insoluble in subsequent cleaning steps are removed. (If the history of the wafer is known, this degreasing operation may often be safely omitted.) Wafers are then sent through a series of solutions designed to remove any trace of metals or other potentially harmful materials. A common series of cleaning steps is:

STEP	REASONS
Heat in H_2SO_4.	Removes any photoresist or other organic material.
Heat in aqua regia.	Dissolves gold as well as other metals.
Dip briefly in dilute HF.	Top layer of SiO_2 containing any potential contamination is etched away.
Rinse in water.	Remove any residual acid.
Dry.	Get wafers ready for the next process step.

Table 11–2: Semiconductor Equipment Particle Generators

TYPE	PROBLEM
Gas Distribution and Control Systems	Internal welds, seals, solenoid valves, gas filter media, connectors
Spin Dryers	Drive-mechanism feed-throughs, static charge, exhaust, door seals, loading and handling
Resist Spinners	Exhaust control, splashback, edge lip, plumbing lines, resist bottle or package, dispensing pump
Ion Implantation	Mechanical transport, resist chips and residues, gas ports, vacuum components
Low-Pressure Deposition (LPCVD)	Chemical gas contamination, quartz tube particles, reaction chamber spalling, loading stations
Metal Deposition	Loading mechanisms, internal coating and spalling, source control, internal moving parts
Dry Etching Systems	Autoloaders, pedestals, ionized contamination, sidewall accumulations, internal mechanisms, back-side wafer residues
Annealing Systems	Metallic contamination, loading and transfer systems, internal stress spalling, wafer residue transfer, annealing gas inlets

The chemicals used to remove trace amounts of harmful elements must themselves contain sufficiently low amounts of each of these elements.

In addition to cleaning the wafers that are to be processed, all fabrication equipment that comes in contact with the wafers or is directly linked to wafer processing must be similarly cleaned. This list of equipment includes diffusion tubes and associated glassware, wafer boats, push rods, thermocouple shields, vacuum wands, and all of the processing equipment. Table 11–2 lists various types of processing equipment along with known sources of particulate generation. The cleaning steps must remove these particles to be effective.

11.2 Water

Water is used in the rinsing step at the end of almost every cleaning operation. The frequent use of water makes it imperative that it contain minimum amounts of potentially harmful contaminants. Types of contaminants that may exist even in pure drinking water which cannot be tolerated in water used for microelectronics include:

1. Dissolved inorganic salts such as sodium and calcium salts. These are dissolved by the water as it flows through pipes, rocks, soil, etc.

11.2 Water

2. Dissolved organic compounds from industrial waste or living matter.
3. Particulate matter such as small silica particles from rock, soil, and paper.
4. Microbiological life that sustains itself on other contaminants.

Water containing these types of impurities must be purified to reach the levels in Table 11–3.

The process of *ion exchange*, or deionization, was used almost exclusively in water purification prior to the last few years. Ion exchange is the removal of positive and negative ions using activated resins. A typical ion exchange water system is shown in Figure 11–2. Such a system contains the following elements:

1. *Chemical treatment (often chlorination)*. Kills organisms present in the feed water.
2. *Sand filter*. Removes particles from the incoming water.
3. *Activated charcoal filter*. Removes free chlorine and traces of organic matter.
4. *Diatomaceous earth filter*. Retains additional contaminants.
5. *Anion exchange*. Removes strongly ionized acids such as sulfuric, hydrochloric, and nitric acid.
6. *Mixed bed polisher*. Contains both cation and anion resins, and removes any ions missed by previous exchange filters.
7. *Sterilization*. Controls growth of bacteria; often achieved by chlorination or ultraviolet light.
8. *Filter*. Removes any residual particles left in the wafer prior to its first use.

Systems like the one shown in Figure 11–2 have been partially displaced by *reverse osmosis*, or RO, systems. Under pressure, and in the presence of a selectively permeable membrane, water will flow through the membrane, while

Table 11–3: High-Purity Semiconductor-Manufacturing Water Compared to Tap Water

WATER SPECIFICATION	TAP WATER	HIGH-PURITY WATER
Resistivity (megohm-cm)	.0002	15–18
Electrolytes (parts per billion)	200,000	<25
Particulate content (No./cm^3)	100,000	<150
Living organisms (No./cm^3)	100–10,000	<10

180　　11:　The Wafer Fabrication Environment

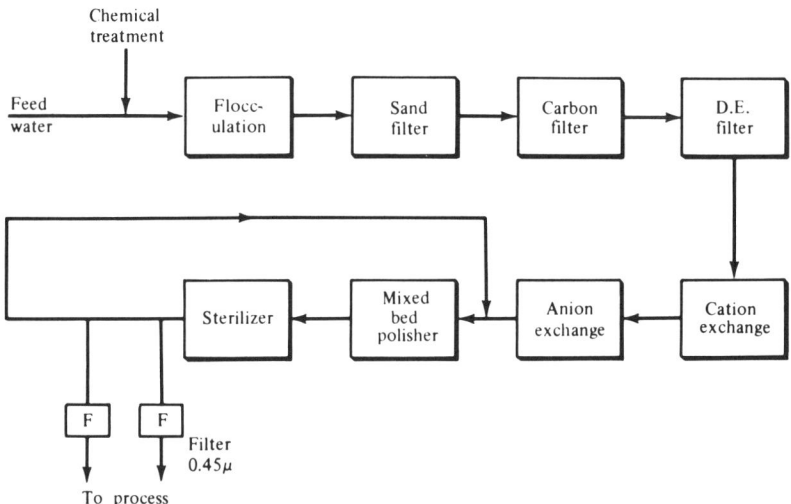

Figure 11–2: Typical ion exchange high-purity water system.

dissolved or suspended substances will not. A diagram of a typical reverse osmosis water system is shown in Figure 11–3.

The differences between this system and the ion exchange system are the pH adjustment, the filtering, and the use of reverse osmosis. All the other steps are identical. Reverse osmosis is effective because it reduces the frequency with which the ion exchange resins must be regenerated.

Once water has been purified to the required level, it is necessary to distribute it throughout the facility without suffering losses in water quality. This

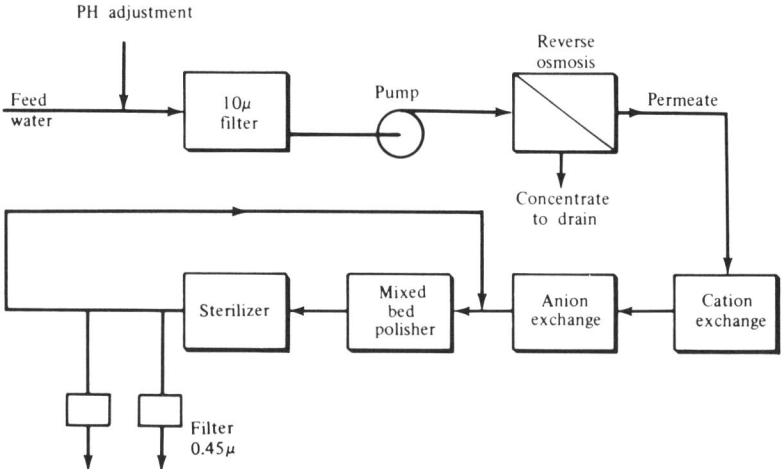

Figure 11–3: Typical reverse osmosis high-purity water system.

11.3 Air

requirement is usually met by distributing the water in a type of inert plastic piping to prevent any significant recontamination. The distribution piping is designed with continuous loops, so the water constantly circulates through the pipes and back through the mixed bed polisher and the sterilizer. Prior to use, it passes through a final filter.

11.3 | Air

The three major parameters that must be controlled in the ambient air are temperature, humidity, and particle level. Temperature and humidity, which affect process steps—particularly photoresist steps—are determined by the type, amount, and set point of the air-handling equipment, but the presence

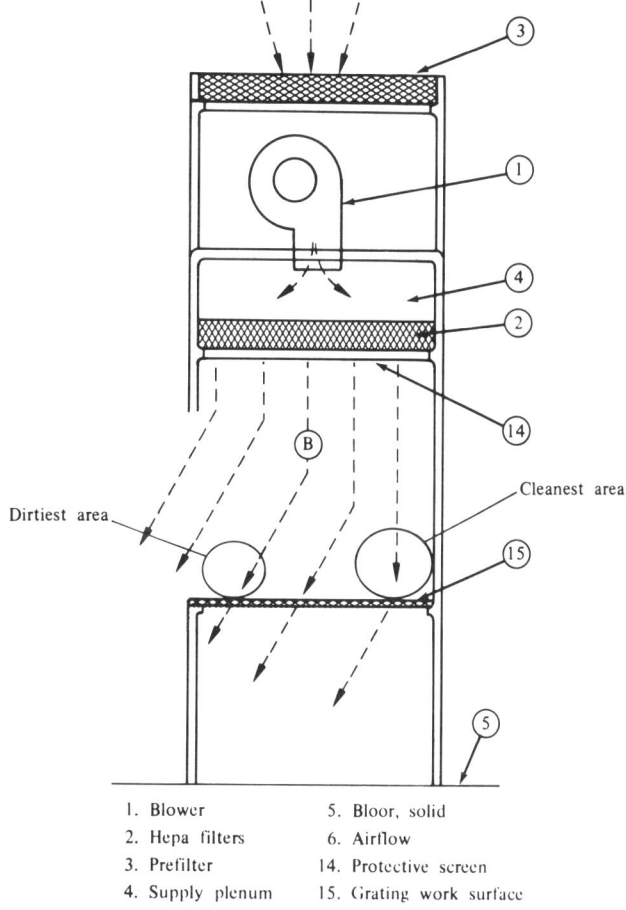

1. Blower
2. Hepa filters
3. Prefilter
4. Supply plenum
5. Bloor, solid
6. Airflow
14. Protective screen
15. Grating work surface

Figure 11-4: Vertical laminar flow hood.

of particles must be controlled in another fashion. Particles prove to be more detrimental to the process at steps such as cleaning and loading, or during the photoresist process. Accordingly, if efforts are made to control the presence of unwanted airborne particles at the critical operations, the process will not suffer. This philosophy brought about the use of laminar flow hoods or stations at certain operations in device fabrication. The cross section of a vertical laminar flow hood is shown in Figure 11-4.

A laminar flow uses the room as a reservoir, taking air from it, filtering the air, and blowing it into the work area in a parallel or laminar flow pattern. Laminar flow prevents the formation of turbulent regions that may accumulate high levels of contamination. The most important part of the laminar flow hood is the *high-efficiency particulate air*, or HEPA, filter. A HEPA filter is a fragile but effective filter capable of removing a minimum of 99.97% of all .2µ or larger particles. Use of tested HEPA filters in all laminar flow hoods is a necessity.

11.4 | Gases

The gases needed in semiconductor fabrication are nitrogen, oxygen, hydrogen, and a variety of source or dopant gases. Precautions must be taken to ensure that no contamination reaches the wafers from these gases. Copper tubing may be used for oxygen and nitrogen, although stainless steel prevents the occurrence of certain kinds of problems. Stainless steel tubing should be used for all other gases because of its high resistance to corrosion. Filters are an important item and are installed near the point of use of the water to prevent the passage of any unwanted particles.

11.5 | Particle-Monitoring Technology

A variety of particle-monitoring techniques are available to specifically detect the multitude of defects and contamination problems associated with the semiconductor manufacturing process. Table 11-4 lists the various monitoring methods along with the detection limits of each technology.

Automated surface inspection systems using either visible light or laser beams will identify the defects and contamination on the wafer surface. Figure 11-5 shows a schematic representation of a typical laser inspection system. Histograms by particle diameter are also available down to the 0.3-micron particle size.

11.6 Personnel and Clean Room Procedures

Table 11–4: Monitoring Technology

TYPE OF MONITOR	TECHNOLOGY USED	LOWER SENSITIVITY LEVEL (MICRONS)
Surface	White light	0.5
Surface	Laser	0.3
Aerosol	White light	0.3
Aerosol	Laser	0.1
Aerosol	Condensation	0.01
Liquid	White light	0.5
Liquid	Laser	0.4
Liquid	Ultrasonics	1.0

11.6 Personnel and Clean Room Procedures

The greatest sources of contamination in a wafer fabrication area are the people that perform the operations. Human beings are continuous sources of organic matter because of the constantly renewing nature of their bodies. It is possible to neutralize the presence of operators by instituting certain sets of procedures. The first step is to partition off the wafer fabrication area. Smocks for operators were used in the past, but they do little more than serve as a

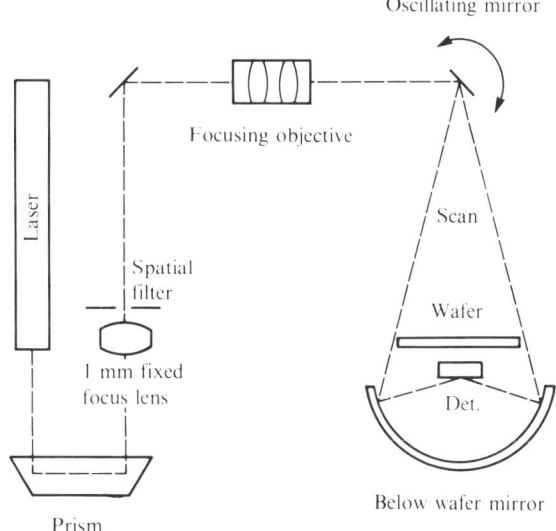

Figure 11–5: Scanning laser technology for surface particle detection.

protective layer. Their impact on contamination control is minimal. The most effective method of controlling contamination from operators is to provide them with contamination-free clothing that covers as much of their body as possible, including an outer covering on the hands and the feet. These suits must be cleaned regularly to keep the level of accumulated contamination low.

Review Exercises

1. What chemical, present in the human body, is a mobile contaminant in silicon dioxide?
2. State two methods of obtaining high-purity water for semiconductor processing.
3. For wafers of unknown origin, should an acid cleaning or a solvent cleaning be first? Why?
4. Why are laminar flow hoods placed over critical areas in a wafer fabrication area? Would the area benefit if the ceiling was laminar flow?
5. List and describe the common cleaning steps utilized to prepare wafers for operations in the device fabrication sequence.
6. Sketch and label a typical ion exchange water purification system.
7. List the major differences between the ion exchange and reverse osmosis water purification systems.
8. How is purified water distributed to various points within a semiconductor processing facility?
9. What metal or metals are frequently utilized to transport the various gases required during semiconductor processing?
10. Sketch and describe a typical wafer fabrication area, focusing on contamination control.
11. Graph the yield equation 11–1 for $N = 9$ and $D = 5$ defects per square inch. How much does the yield decrease for a die that is 100 mils on each side when the defect density per square inch increases from 2 to 5?

12 Semiconductor measurements

12.0 Introduction

Throughout the semiconductor manufacturing process, the result of each major processing step must be carefully evaluated. This chapter explores the fundamental testing methods used to evaluate the basic steps of crystal growth, epitaxial deposition, oxidation, ion implantation, diffusion, and metallization.

12.1 Evaluation of Semiconductor Materials

The amount of dopant in a wafer, bar, or sample of doped silicon is determined by first measuring the sheet resistance R_S of the wafer, bar, or sample, in ohms per square ($\Omega/\square$). Sheet resistance is very often measured using an in-line four-point probe, as illustrated in Figure 12–1. The formula relating sheet resistance to the impressed current I and measured voltage V is

$$R_S = 4.53\left(\frac{V}{I}\right) \tag{12-1}$$

The probe spacing s in the figure is typically either 1000 micrometers (40 mils) or 1250 micrometers (50 mils) and is very precise. The constant in equation 12–1 is valid only for a two-dimensional sheet of material extending to infinity in both directions. A correction factor must be included for noninfinite sam-

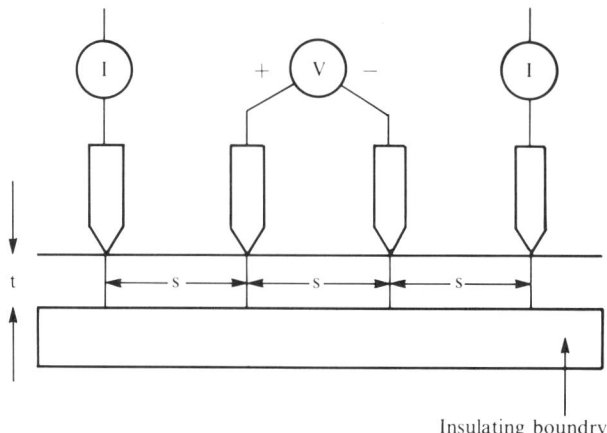

Figure 12–1: An in-line four-point probe.

ples which depends upon the sample geometry and the probe spacing. Several corrections must be made to equation 12–1 if the sample is a thick material or has a small, finite surface area.

First, consider a thick sample. In general, if the sample layer is more than 0.6 times the probe spacing, the sheet resistance is reduced by the correction factor. For typical probe spacings of 1000 and 1250 micrometers, this corresponds to samples thicker than 600 micrometers (24 mils) and 750 micrometers (30 mils). Equation 12–1 must accordingly be multiplied by the correction factor A, which is a function of the ratio of the film thickness t to the probe spacing s. This relationship is illustrated in Figure 12–2, and the resulting equation is:

$$R_S = 4.53\left(\frac{V}{I}\right)A \tag{12-2}$$

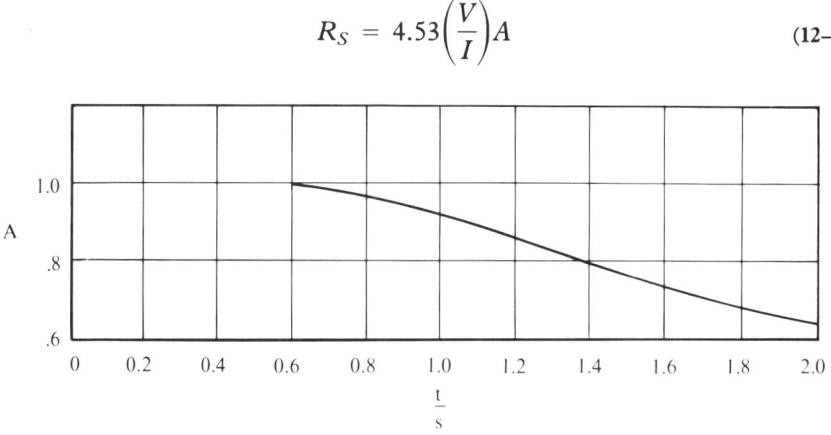

Figure 12–2: Correction factor for thick materials on a four-point probe.

12.1 Evaluation of Semiconductor Materials

Next, consider a sample that does not extend to infinity in both directions but rather is of small size. The sheet resistance measurement for a small sample is influenced by edge effects. Two geometries frequently encountered are those of the rectangle and the circle, as illustrated in Figure 12–3.

For these small geometries, the correction factor B is a function of the ratio of the dimension D to the probe spacing s. This is illustrated in Figure 12–4.

For a thick, small sample, the correction factor is simply the product of the two individual correction factors.

If the layer is uniformly doped, the sheet resistance may be converted to resistivity by simply multiplying by the thickness of the layer, yielding

$$\rho = R_S \cdot t \tag{12-3}$$

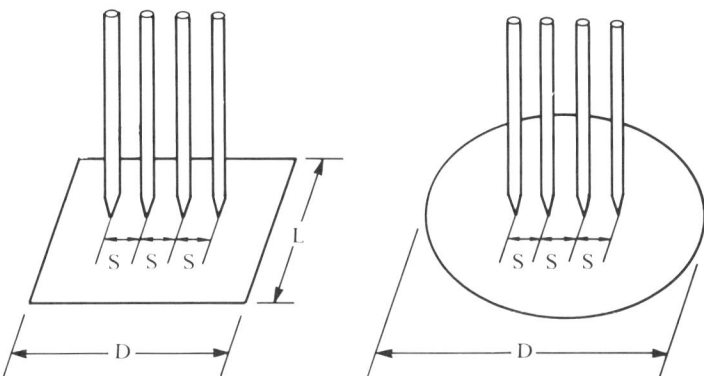

Figure 12–3: Rectangular and circular small sample geometries.

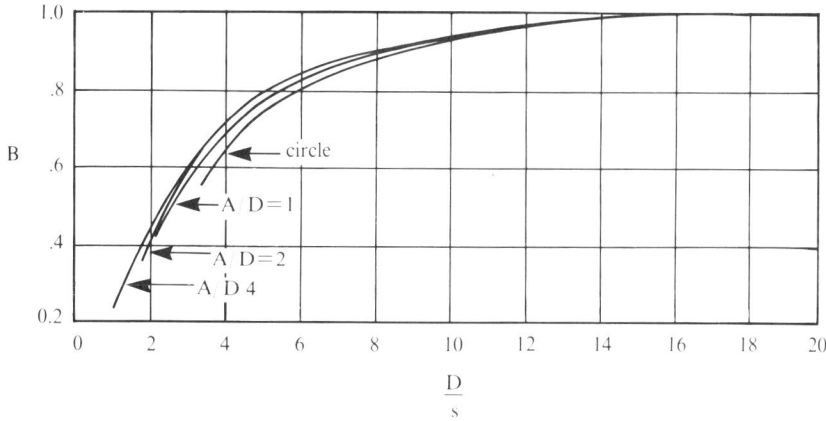

Figure 12–4: Correction factors for rectangular and circular small-sample geometries.

For nonuniformly doped diffused layers, only the average resistivity may be calculated. Common practice in the semiconductor industry is to define the average resistivity in terms of the junction depth x_j, as follows:

$$\rho_{\text{avg}} = R_s \cdot x_j \quad (12\text{--}4)$$

12.2 Evaluation of Diffused Layers

As mentioned in the previous section, diffusion results can be checked by a measurement of sheet resistance and junction depth to give an average resistivity of the diffused layer. The junction depth x_j (measured in micrometers) is defined as the distance from the top surface to the depth within the diffused layer at which the dopant concentration equals the background concentration. This depth is commonly measured by first chemically staining a beveled (1 to 5°) sample within the diffused region. The smaller angle is used for the shallow diffused junction in order to obtain better spatial resolution. The bevel is made by using an angle-lap fixture as shown in Figure 12–5.

The sample is stained with a mixture of 100 milliliters of hydrofluoric acid (49% dilution) to which a small amount of nitric acid has been added. Hydrofluoric acid alone is sometimes used. The stain is swabbed onto the surface under strong illumination for up to two minutes, after which the p-type regions are darker than the n-type regions. Once stained, the sample is observed under collimated, monochromatic light, as shown in Figure 12–6. An interference pattern will be projected onto a cover glass, and the junction depth may be calculated by counting the interference fringes and then applying the equation

$$x_j = n\left(\frac{\lambda}{2}\right) \quad (12\text{--}5)$$

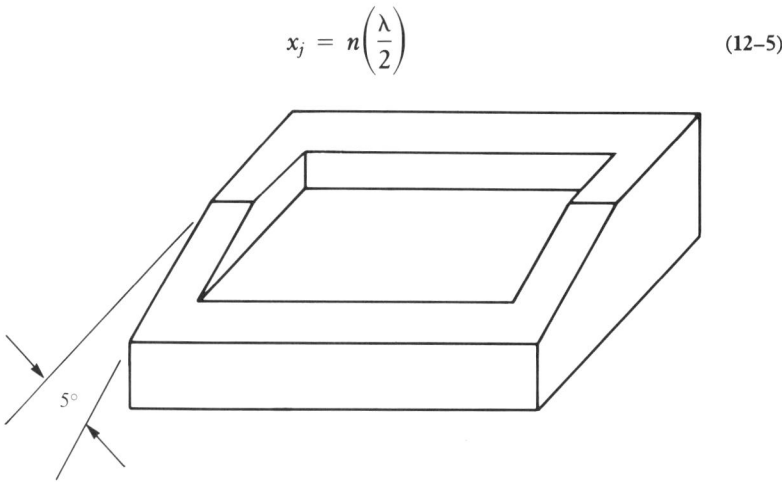

Figure 12–5: A 5° angle-lap fixture.

12.3 Diffusion Profile Measurements

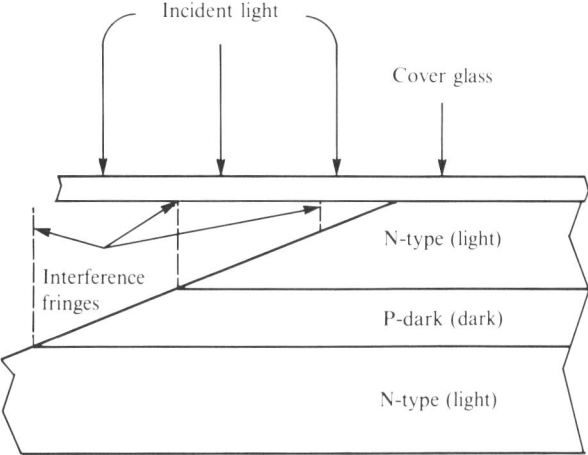

Figure 12–6: Lap-and-stain measurement set-up.

where

n = the number of fringes

λ = the wavelength of the incident illumination.

If the diffusion profile is known (e.g., erfc or Gaussian), the average resistivity ρ_{avg}, as calculated by Equation 12–4, is uniquely related to the surface concentration of the diffused layer and the substrate dopant concentration. Design curves, sometimes called *Irwin's curves* after the original data presented by Irwin at Bell Labs, have been calculated for the more common exponential Gaussian and erfc distributions. Unfortunately, the diffusion profiles cannot be represented by these simple profiles for high dopant concentrations and shallow junctions. However, because of their relative simplicity and ease of use, the Gaussian and erfc profiles are still used by treating the resulting parameters as effective values from which the process engineer can tune the diffusion process.

12.3 Diffusion Profile Measurements

The simple measurements of junction depth and sheet resistance are useful as process-monitoring tools but are inadequate for detailed diffusion analysis. For more in-depth diffusion study, the process engineer must turn to more accurate profiling tools such as capacitance–voltage (CV) measurement or spreading-resistance analysis.

CV profiling utilizes the measurement of the decrease in capacitance of a reverse-biased depletion layer as this layer moves farther into the semiconductor. Because the extent of the depletion layer depends on the impurity concentration, the capacitance in this case depends on capacitor geometry (e.g., plate area), the number of carriers, and the bias voltage, as given by the equation

$$\left(\frac{C_i}{C}\right)^2 = \frac{2C_i^2}{\epsilon_s qN}(V_{bi} - \phi) + 1 \tag{12-6}$$

where

$N =$ the doping concentration

$\phi =$ the built-in potential of the junction

$C_i =$ the capacitance due to the initial geometry, ϵ_s is the permittivity of SiO_2 and q is the electronic charge (1.6×10^{-19} coulombs).

If the capacitance is measured and plotted as a function of bias voltage V_{bi}, then the doping concentration N can be determined to within a few diffusion lengths (usually a few tenths of a micron) of the surface.

The CV measurement technique is generally limited to distances of below a few diffusion lengths of the surface and dopant concentrations between 10^{13} and 10^{17} atoms/cm^3. For these reasons, an alternate technique of spreading-resistance analysis has been developed for diffusion profile measurement.

As shown in Figure 12-7, the resistance of a wire with radius r and length l is given by

$$R = \frac{\rho \ell}{\pi r^2} \tag{12-7}$$

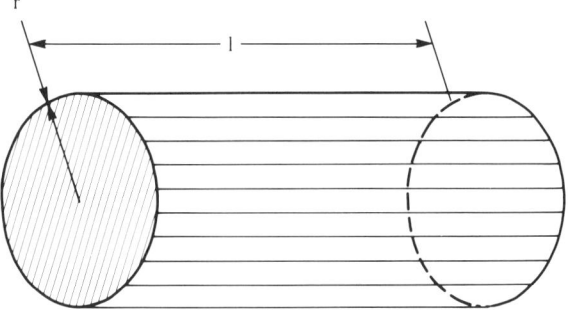

Figure 12-7: Wire cross section with uniform current flow.

12.3 Diffusion Profile Measurements

where ρ is the resistivity of the wire material. Now, suppose that the size of the contacts at the ends of the wire are reduced so that they no longer cover the entire end surface of the wire. Then, as their size is decreased, the resistance between the contacts will increase, and the current will no longer flow uniformly (parallel to the wire axis). Such a condition is illustrated in Figure 12–8. For this case, the resistance is determined solely by the resistivity and contact radius a, as given by the equation

$$R_{sr} = \frac{\rho}{2a} \tag{12-8}$$

where

R_{sr} = the spreading resistance

ρ = the average resistivity near the contact point.

Note that the two contacts need not be on opposite ends of the wire, but can be side by side on the same surface. Such an arrangement is illustrated in Figure 12–9.

The contacts used for spreading-resistance measurements on silicon are simply alloy probe tips under pressure (about 10 grams) against the surface. Before the measurements, the sample is beveled in a manner much like that used for junction-depth measurement, so that the probes may be stepped along the exposed diffused layer. The two probes are spaced as close as possible (10 to 100 micrometers), and measurements are made at various distances from the bevel edge. The depth below the original surface is calculated for each measured point. Bevel angles anywhere from 0.1 to 10° are used, depending upon the required resolution.

The profiles provided are not those of the actual dopant concentration versus depth, but rather, those of the net carrier concentration versus depth. The resistivity is obtained from the measured spreading resistance and the

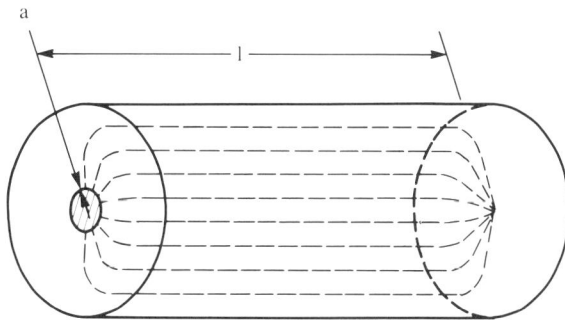

Figure 12–8: Wire cross section with contact radius a>r.

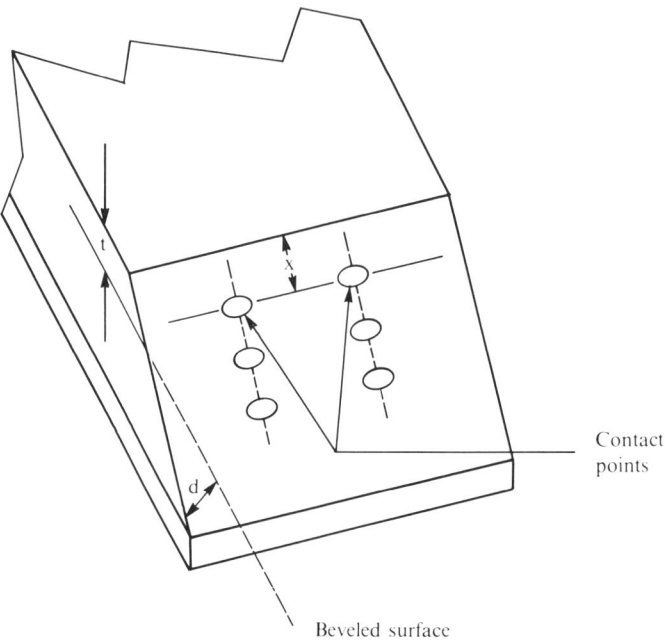

Figure 12–9: Beveled sample showing probe contact locations.

probe calibration curve of spreading resistance versus sample resistivity, rather than from equation 12–8. Once the resistivity is known, the carrier concentration at each point probed along the bevel can be calculated from the equation

$$N = \frac{1}{\rho q \mu} \tag{12-9}$$

where

q = the electronic charge

μ = the bulk mobility appropriate to each resistivity measurement.

Table 12–1: Diffusion Profile Measurement Methods

METHOD	RANGE/LIMITS	COMMENTS
Capacitance–Voltage	2×10^{12} atoms/cm^2	Poor surface resolution
Spreading Resistance	>1 μm depth	Beveled sample
SIMS	0.1 μm depth	Requires standard
Differential Conductivity	10^{18} to 10^{20} atoms/cm^3	Requires surface etch

12.4 Epitaxial Layer Evaluation

Various other techniques, such as differential conductivity and the secondary ion mass spectroscopy (SIMS), have been used for impurity profile measurements. Table 12–1 lists the commonly used diffusion profile measurements, along with their characteristics and limitations.

12.4 Epitaxial Layer Evaluation

The epitaxial layer parameters of most interest to the process engineer are the layer doping and thickness. In addition, the performance of the circuits or devices which are constructed within this epitaxial layer depends upon the density and distribution of crystal defects within the layer.

Epitaxial layer thickness can be measured using the technique of groove and stain (or angle lap and stain), measuring etch pit depth, or infrared interference. The first two methods are covered in detail in Chapter 6. The third method has become increasingly popular in recent years through the use of the infrared spectrophotometer. In addition to providing a rapid and accurate measurement of epitaxial thickness, this instrument also performs carbon and oxygen analysis.

Figure 12–10 illustrates the basic principle of the method first described by Spitzer and Tanenbaum to measure epitaxial thickness by reflectance. The incoming infrared beam is focused on the surface of the epitaxial layer at an angle (usually about 15°) at which the beam is split by reflection and refraction. The refracted ray continues on through the epitaxial layer, where it is

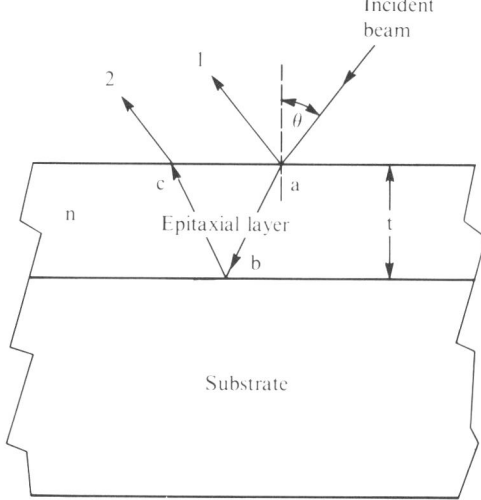

Figure 12–10: Interface reflections through an epitaxial layer.

reflected at the epi-substrate interface (point b). The actual reflection point is at the heavily doped point on the outdiffusion profile, which for most common processing conditions is usually near the interface. Depending on the optical path difference between the reflected ray at point a and the emerging ray at point c, constructive or destructive interference will occur. A minimum will occur when the path difference is an even multiple of half wavelengths, and a maximum will occur when the difference is an odd multiple of half wavelengths. The optical path difference is a function of θ, the angle of incidence; t, the thickness of the layer; n, the index of refraction of the layer; and λ, the wavelength of the incident radiation. Since θ is fixed and n is known, t can be determined by the relationship between the optical path difference and λ. The dependence of thickness on various parameters is given in the equation

$$t = \frac{(N - 1/2 + C_s)W_n}{2(n^2 - \sin^2\theta)} \quad (12\text{--}10)$$

where

N = the order of the maxima or minima

C_s = a substrate correction factor

θ = the angle of incidence

n = the refractive index

W_n = the position, in micrometers, of the maximum or minimum in the spectrum of the material.

The most frequently used methods of determining epitaxial layer impurity profiles are measuring spreading resistance, reverse bias CV plotting, and sectioning. The spreading-resistance technique was covered in considerable detail in Section 12.3. The preferred approach is the use of diode CV measurements. A mercury probe is frequently used to form the contact with the semiconductor surface. Figure 12–11 illustrates how the mercury is drawn through a siphon tube to make contact with the wafer, which is positioned face down. The mercury contact area is first determined by measuring the capacitance of an oxide layer of known thickness.

Mercury in contact with a semiconductor forms a Schottky diode with a voltage-dependent capacitance. For a nonhomogeneous doping concentration, as found in an epitaxial layer, the concentration N is related to the depletion capacitance as given in the equation

$$N(x) = \frac{C^3(V)}{q\epsilon_s \dfrac{dC}{dV}} \quad (12\text{--}11)$$

12.4 Epitaxial Layer Evaluation

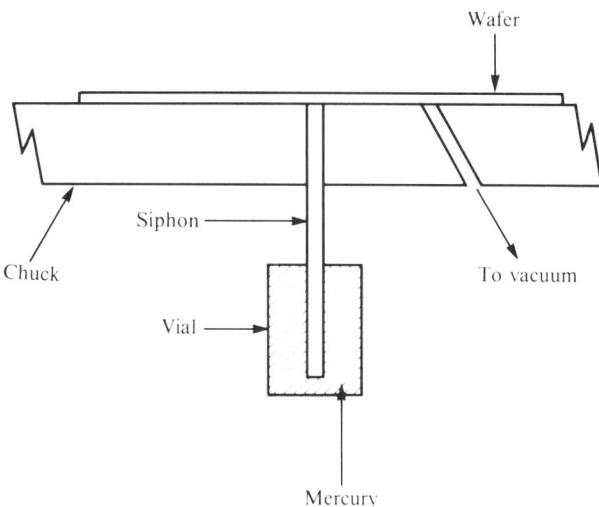

Figure 12-11: Cross section of mercury probe for C-V plotting.

where

$N(x)$ = the impurity concentration at the space-charge layer edge

$C(V)$ = the junction reverse-bias capacitance per unit area

A = the junction area.

The width W of the depletion region is given by

$$W = \frac{\epsilon_s A}{C(V)} \tag{12-12}$$

In CV instruments, the measurements are normally performed on an incremental basis. That is, the voltage is first incremented by a small amount, and then the new capacitance is measured and plotted. Equations 12-11 and 12-12 thus become

$$N = \frac{(C_i + C_{i+1})^3 (V_i - V_{i+1})}{q\epsilon_s(C_{i+1} - C_i)8A^2} \tag{12-13}$$

and

$$W = \frac{\epsilon_s A}{(C_i + C_{i+1})/2} \tag{12-14}$$

where, for two incremental capacitance–voltage readings,

C_i = initial capacitance reading, in farads, at V_i

C_{i+1} = next capacitance reading, in farads, at V_i

V_i = initial voltage applied in volts (reverse bias)
V_{i+1} = next-step voltage applied in volts (reverse bias)
q = 1.6 × 10^{-19} coulombs
ϵ_s = 1.03662 × 10^{-12} farads/meter.

The surface of the epitaxial layer is normally treated by different methods, depending upon the impurity type. When the wafer leaves the epitaxial reactor, it is highly reactive, and the surface will combine with ambient oxygen (to form native oxide) and any contaminants in the surrounding environment. A chemical cleaning and oxidation will usually minimize these effects and leave an exceptionally clean 20- to 25-angstrom oxide layer, which yields a "good" diode characteristic.

Sectioning is probably the most difficult-to-perform method of determining epitaxial layer impurity profiles and is also the least accurate. In it, layers of silicon are removed from the surface of the wafer using anodic oxidation or chemical etching. The sheet resistance of each newly exposed layer is then determined, and the data are mathematically manipulated and plotted to yield impurity concentration as a function of depth into the surface. At least one attempt has been made to automate the process using an electrochemical cell in direct contact with the wafer to perform the etching.

12.5 Oxide and Thin-Film Evaluation Techniques

The three principal areas normally evaluated to determine the results of a thermal oxidation process are thickness, breakdown voltage, and quality (level of contamination). Oxide thickness may be determined by a number of different methods, including color charts, as described in Chapter 3; however, the most prevalent methods used during the manufacturing process are ellipsometry, interferometry, reflectometry, CV, and profilometry. The prism coupler has also found application in the measurement of thin films, including thermally grown oxide.

The measurement of thermally grown oxide up to tens of thousands of angstroms thick is frequently accomplished with ellipsometry. Ellipsometry involves reflecting polarized, monochromatic light from the surface of the film; analyzing the changes in intensity, phase, and polarization of the reflected light; and solving a set of Fresnel equations which were modified by Drude to include films. As shown in Figure 12–12, elliptically polarized light, usually from a linearly polarized helium–neon laser and a compensator, is reflected off the surface of the film to be measured. The reflected wave will be linearly polarized into two components R_p and R_s, with a phase difference between the p and s components of β. Note that the incident laser light contains two electric field components E_p and E_s, with a phase difference between p and s of α.

12.5 Oxide and Thin-Film Evaluation Techniques

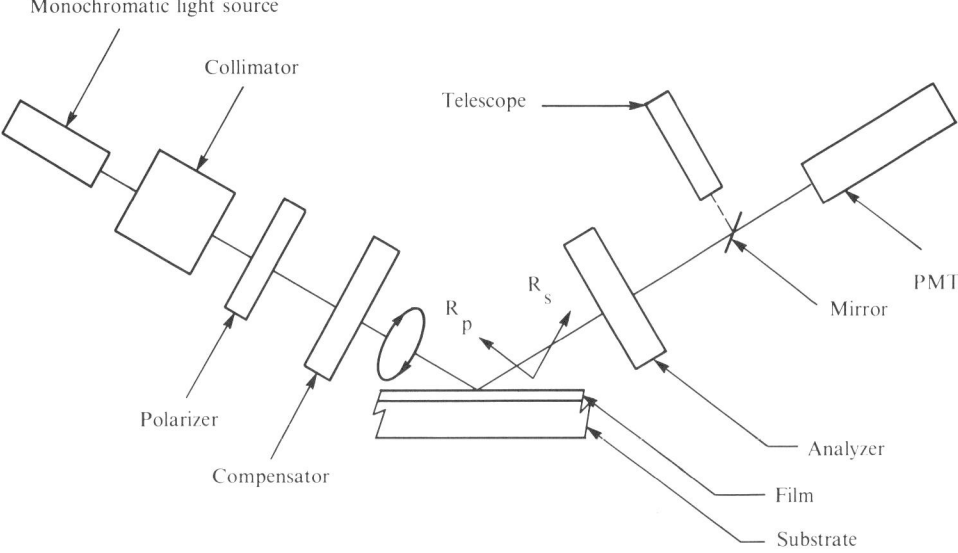

Figure 12–12: Simplified diagram of an ellipsometer.

The ratio of R_p and R_s, the ratio of the reflection coefficients p_p and p_s, and the phase difference β are obtained by analyzing the reflected light. The ratio of E_p to E_s and the phase difference α are known for the instrument. The relationships between these variables are illustrated in the equations

$$\tan \psi = \left(\frac{R_p}{R_s}\right)\left(\frac{E_p}{E_s}\right) \quad (12\text{--}15)$$

and

$$p = \tan \psi \, e^{i\Delta} \quad (12\text{--}16)$$

where

$$p = \frac{p_p}{p_s}$$

and

$$\Delta = \beta - \alpha$$

Ellipsometry, then, involves the measurement of the tangent of the angle ψ, the change in the amplitude ratio p upon reflection, and the change in the phase difference Δ upon reflection. These ratios depend upon the wavelength of the light being used, the angle of incidence of the light, the optical properties of the ambient medium, and most importantly, the thickness and optical constants of the film. If the optical constants of the bare surface are known, the thickness and refractive index of a film on that surface may be calculated from the measured values of Δ and ψ. Considerable computation is required, and unambiguous results are obtained only if it is known in advance that the film lies within a restricted range, usually about zero to 3000 angstroms.

Optical interferometry may also be used to measure oxide film thickness with a polychromatic light source. Figure 12–13 is a schematic diagram of one such device. The spectrum of light reflected from a film over a substrate is characterized by maxima and minima, which for a known wavelength are given by the equation

$$\lambda = \frac{2t_f}{p}(n_f^2 - \sin^2\theta)^{1/2} \tag{12-17}$$

where

t_f = the film thickness

n_f = the refractive index of the film

θ = the angle of incidence

p = an integer based on the order of interference.

The ratio of maximum to minimum reflectivity depends on the refractive indices of both the film and the substrate. When the refractive index of the substrate is greater than that of the film, the reflectivity as a function of wavelength will vary according to Equation 12–17. It is also clear from this equation that unless the refractive index of the film is known, two measurements

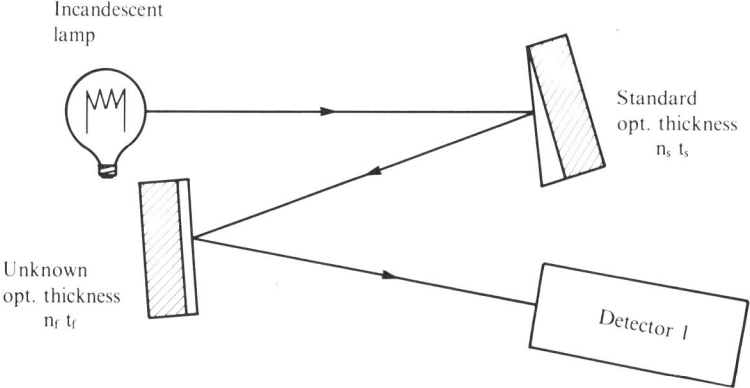

Figure 12–13: Schematic diagram of an optical interferometer.

12.5 Oxide and Thin-Film Evaluation Techniques

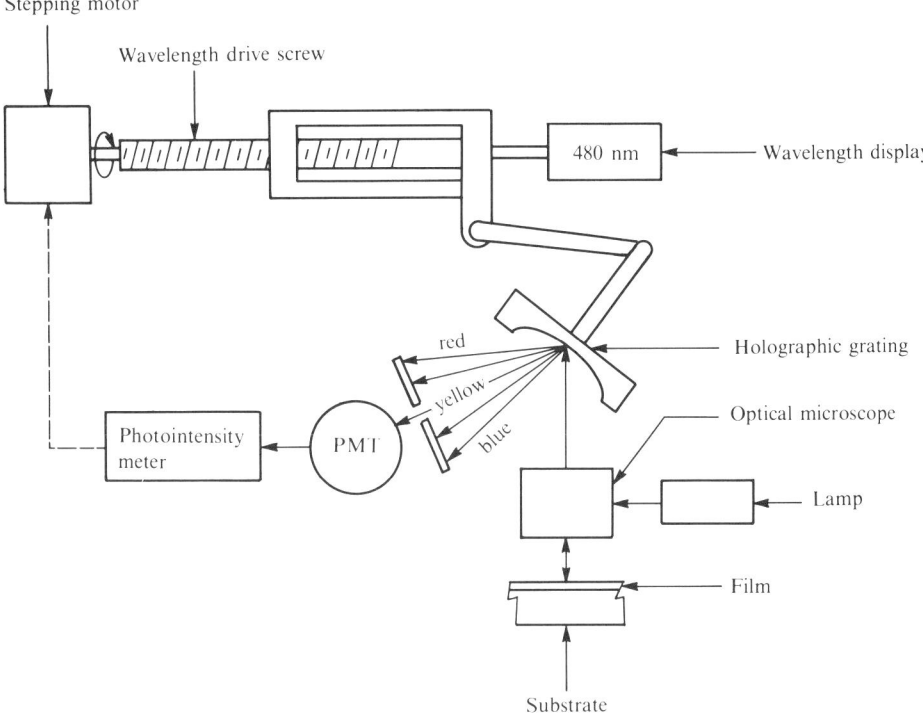

Figure 12–14: Schematic diagram of a microspectrophotometer.

at two different angles of incidence must be made in order to determine n_f. In many cases, it is only necessary to determine the *optical* thickness $n_f t_f$. If, as shown in Figure 12–13, the light from the source is first incident upon a standard film of known optical thickness $n_s t_s$, and then upon the film being measured, a strong maximum will occur when the unknown film has the same optical thickness as the standard film. In order to *scan* the optical thickness and look for this maximum, it is necessary to utilize a variable thickness standard. This may be constructed by etching a silicon dioxide film on a silicon substrate in such a manner that the thickness of the film (in the form of a disk) varies linearly as a function of the angle of rotation of the disk. The disk is then rotated, and the signal at the detector is compared to a portion of the reference signal for a maximum. The position of the disk then yields the optical thickness $n_f t_f$ of the unknown film. If the refractive index of the film is known, the user may calculate the film thickness.

An alternative approach utilizing optical interferometry is illustrated in Figure 12–14. A tungsten light source illuminates the film surface. Interference of the light rays caused by multiple reflections inside the film yields a roughly sinusoidal variation as a function of wavelength. The frequency of the sinusoidal variation is a simple function of the thickness and refractive index of the film. A stepping motor is used to scan a holographic grating and slit over the wavelength range from 480 nanometers (blue) to 800 nanometers (deep red) of the color spectrum.

Both of the aforementioned instruments have the disadvantage of requiring that the user provide the refractive index of the film under test. Several techniques have recently been introduced which are independent of refractive index. The first method involves a reflectance measurement at four different angles of incidence using a polarized helium–neon laser. These angles are controlled by a servomotor which adjusts the tilt of the substrate stage relative to the light beams. Since multiple angles of incidence are now available, it is possible to calculate the dielectric film thickness directly without having to know the refractive index in advance. In addition, the refractive index may also be determined from the instrument.

The second method, prism coupling, involves intimate contact with a prism (GGG or rutile), as illustrated in Figure 12–15. A laser beam is directed at the base of a prism with a high refractive index and is reflected into a photodetector. The film and substrate are then brought into intimate contact with the prism by a spring-loaded plunger. The angle of incidence θ of the laser beam is varied, and at certain discrete values of θ called mode angles, photons tunnel from the base of the prism into the film and enter into so-called optical propagation modes. When these modes are entered, the intensity of the light reflected into the photodetector drops sharply. If two or more mode angles are observed, it is possible to calculate both the film thickness and refractive index. The disadvantage of the instrument is that it requires contact between the prism and the film.

The most straightforward technique for measuring film thickness requires etching a step or masking off part of the substrate during deposition. A stylus is then drawn over the step thus created, and its vertical position is electroni-

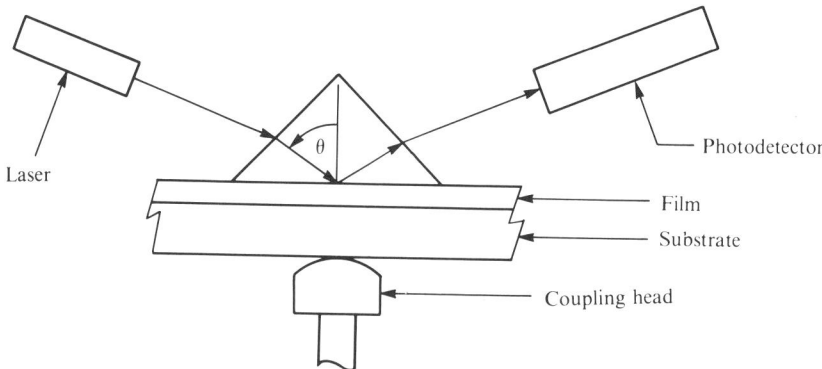

Figure 12–15: Diagram of the prism coupling method.

12.5 Oxide and Thin-Film Evaluation Techniques

cally determined. Accurate step-height measurements as small as 100 angstroms are routinely measured by this method.

The CV plotter may also be used to determine thermal oxide thickness. The thickness t_f of the oxide layer is determined from the measurement of the maximum capacitance C_{max} in Figure 12–16, as given by the equation

$$t_f = 1.72 \times 10^8 \frac{D^2}{C_{max}} \text{ angstroms} \qquad (12\text{--}18)$$

where

D = the diameter of the contact in inches,

C_{max} = the maximum capacitance, measured in picofarads,

t_f = the film thickness, measured in angstroms.

Table 12–2 lists the characteristics of several of the more common film thickness measurement instruments. It should be noted that a wide range of film types besides thermal oxide can also be measured.

In addition to measuring thickness, the CV plotter can be used to measure the amount of mobile ion (typically sodium) contamination in a grown oxide film. The level of contamination is indicative of the condition of the furnace that was used to deposit the film.

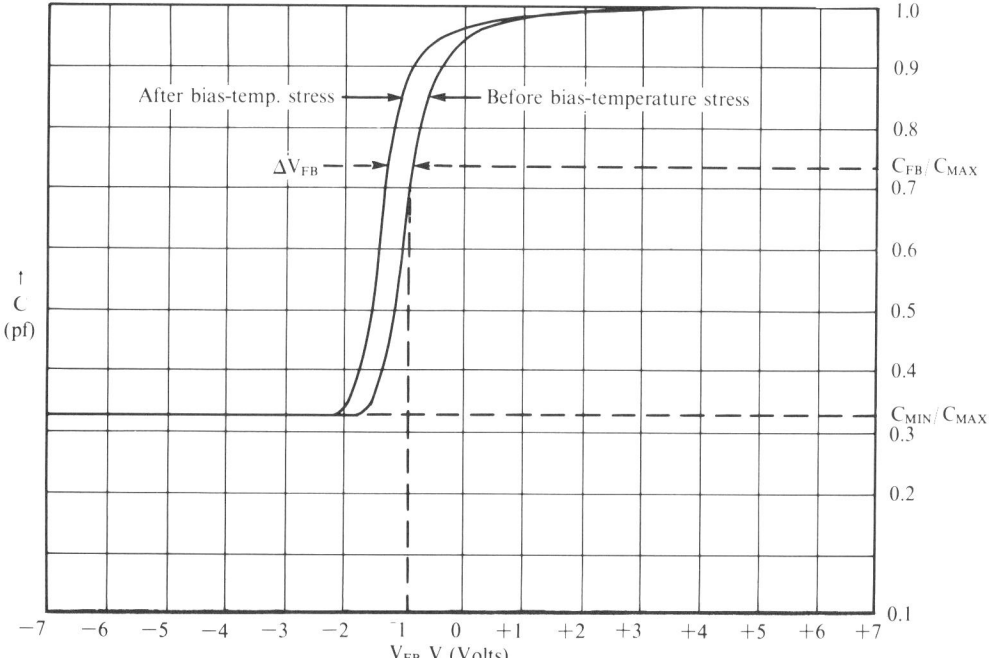

Figure 12–16: Typical C-V plot of MOS capacitor on *n*-type substrate.

Table 12–2: Film Thickness Measurement Instruments

	ELLIPSOMETER	REFLECTOMETER	SPECTROPHOTOMETER	INTERFEROMETER	PRISM COUPLER
Measurement Area	50μ × 50μ	50μ × 125μ	3.5μ–35μ	2000μ × 6000μ	1000μ × 1000μ
Thickness[1] Range	25Å–3KÅ	100Å–20KÅ	100Å–30KÅ	2KÅ–50KÅ	2KÅ–30KÅ
Accuracy	±10Å	±10Å or 0.1%	±20Å or 2–5%	±150Å	±50Å
Index of Refraction	N/A[2]	1.4–3.7	N/A	N/A	1.4–3.8
Minimum Film Thickness					
SiO_2	25Å	100Å	100Å	2000Å	2000Å
Photoresist	50Å	100Å	500Å	—	2800Å
Nitride	25Å	100Å	100Å	1500Å	800Å
Polysilicon	—	100Å	400Å	—	—
Multilayer films	yes	yes	yes	no	yes

[1]Depends upon film type.
[2]N/A = not applicable.

In CV plotting, an MOS capacitor is measured, producing a graph of capacitance versus voltage. After the initial CV plot is made, the capacitor is bias-temperature stressed. This is done by applying a positive voltage (typically +100 volts) to the MOS capacitor and activating the controller to the hot chuck holding the wafer. The controller brings the hot chuck and, hence, the wafer to a preset temperature (usually 300°C) within a 5-minute period. The wafer is held at this temperature for 3 to 10 minutes with the bias applied to the device. At the end of the stress cycle, the wafer is brought back to room temperature by running coolant through the chuck. Finally, a second CV plot is completed, and the two plots on the same paper are compared.

Figure 12–16 illustrates typical (normalized) CV plots for the MOS capacitor before and after exposure to a high-temperature cycle with positive bias. The effect of the bias-temperature stress is to drive any mobile ions in the oxide layer from the region near the metal contact to the boundary of the semiconductor surface. The additional charge at the semiconductor surface created by these mobile ions causes a shift in the CV plot along the voltage axis in the negative direction. The number of mobile ions per unit area in the oxide can be calculated from this voltage shift using the equation

$$N_I = 1.2 \times 10^6 \frac{C_{max} \Delta V_{FB}}{D^2} \text{ ions/cm}^2 \quad (12\text{–}19)$$

where

D = the diameter of the metal contact in inches.

C_{max} = the measured maximum capacitance in picofarads,

ΔV_{FB} = the voltage shift in the flatband voltage.

Mobile ion contamination in oxide—particularly sodium—can cause a shift in the operating parameters of MOS devices. For this reason, CV bias-temperature stress measurements are an important production technique of screening fabrication processes.

A second measure of dielectric quality is the breakdown strength of the oxide layer. Measuring this is a simple task, performed by increasing the voltage across the dielectric until the current rises abruptly. For silicon dioxide, 600 V/μm is considered an acceptable dielectric strength.

12.6 | Ion Implant Evaluation

The commonly used methods to evaluate ion dose and dose variations are the resistive techniques (e.g., four-point probe), the capacitance method (CV plot), test structures such as the Van der Pauw pattern, spectrometry methods, and most recently, high-resolution optical dosimetry. Of all these techniques, the most frequently used are the automated-mapping four-point probe and the mapping optical dosimeter. The former is simply an automated version of the technique discussed earlier in this chapter. A four-point probe is stepped back and forth across the wafer, and each measurement is stored for subsequent manipulation and display. Such instruments produce equivalue contour maps, which indicate the uniformity and level of the dose across the entire wafer. Correction factors must be applied, however, when the distance to the edge of the wafer is less than ten probe spacings away.

Optical dosimetry has also been used as a technique for monitoring ion implant uniformity and dose. A transparent glass "wafer" is spin coated with a layer of thin positive photoresist. The photoresist is then decolored by exposure to UV annealing, thus becoming transparent. The glass wafer is then placed in the ion implanter and exposed to the energetic beam. Finally, the wafer is removed and placed in an optical dosimeter, as illustrated in Figure 12-17. This instrument measures the transparency of the wafer at multiple points before and after the implant. It calculates the before and after difference in optical density and, knowing the characteristics of the photoresist, determines the implant dose and plots the uniformity.

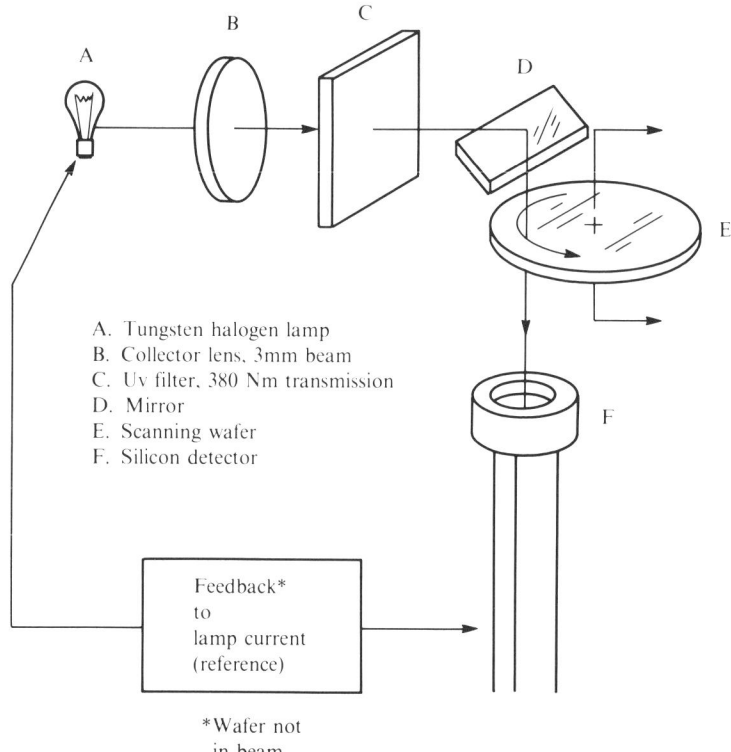

A. Tungsten halogen lamp
B. Collector lens, 3mm beam
C. Uv filter, 380 Nm transmission
D. Mirror
E. Scanning wafer
F. Silicon detector

Feedback* to lamp current (reference)

*Wafer not in beam

Figure 12-17: Optical system diagram of an optical dosimeter.

12.7 Metallization Monitoring

Control of conductive film thickness is essential to the production of high-yield VLSI circuits. Film thickness is monitored both during and after deposition.

The most frequently used in-situ thickness monitor is the quartz crystal resonator plate. The plate is oriented in the deposition system such that the additional mass of the film deposited on the resonator causes frequency changes that can easily be measured. After calibration, the monitor resonator can be used to control the final film thickness, as well as the deposition rate. The monitor is easily calibrated by measuring the increase in mass Δm during a deposition. This increase, together with the area A of the film and the bulk density p_D of the material, can be used to determine the film thickness from the equation

$$t = \frac{\Delta m}{p_D A} \qquad (12\text{--}20)$$

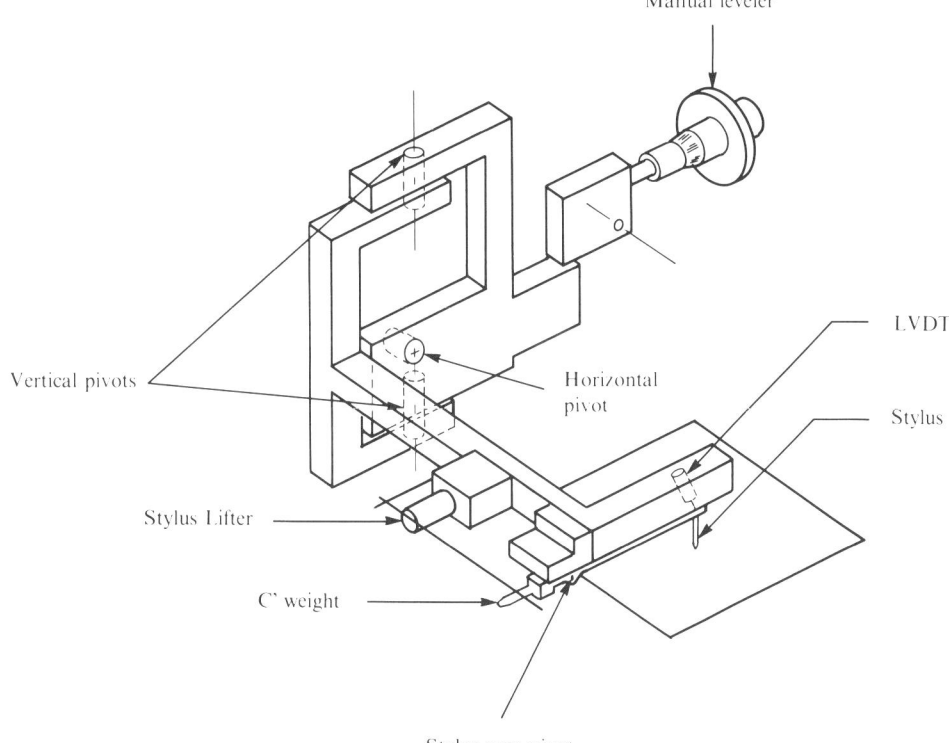

Figure 12–18: Scanner portion of stylus profilometer.

Another technique for measuring the thickness of the film after deposition is to use a surface profilometer. A portion of the film is first etched away down to the surface of the substrate, and a fine diamond stylus is drawn over the surface of the substrate until the etched step is encountered. Figure 12–18 illustrates the scanner portion of such an instrument, in which the vertical motion of the stylus is detected by a linear variable differential transformer. Surface profilometers can measure films (step heights) as thin as 100 angstroms or less.

Review Exercises

1. Calculate the number of mobile ions per unit area for the CV plot of Figure 12–16 for a 0.040-inch-diameter contact if C_{max} is 277 picofarads.
2. Calculate the thickness of the oxide layer for the plot in problem 1.
3. Determine the sheet resistance of a thin, uniformly doped layer of silicon if the probe spacing of the four-point probe equals the silicon thickness and $V/I = 45$.
4. Calculate the resistivity of the sample if the thickness of the silicon layer in problem 3 is 12 micrometers thick.
5. List four commonly used diffusion profile measurements.
6. Determine the flatband voltage and flatband capacitance for the MOS capacitor structure tested in problem 1.

REFERENCES (CHAPTER 12)

1. Spitzer, W. G., and M. Tanenbaum, "Interference Method for Measuring the Thickness of Epitaxially Grown Films," *Journal of Applied Physics*, 32, 744, 1961.
2. Adams, A. C., D. P. Schinke, and C. D. Capio, "An Evaluation of the Prism Coupler for Measuring the Thickness and Refractive Index of Dielectric Films on Silicon Substrates," *Journal of the Electrochemical Society*, 126, 1539, 1979.

13 | Advanced silicon technology

13.0 | Introduction

Silicon technology emerged as the dominant force in solid-state devices in the early 1960s and has continued to progress at a rapid rate. This progress has resulted from better understanding of the materials, the devices, and the processing steps involved in device fabrication. The net result of this tremendous progress in silicon technology has been an increase in the types of functions that can be performed with silicon devices, coupled with the decrease in the cost per function. Silicon technology seems likely to continue in the manner it has over the next few decades, with a steady progression of advances along a number of fronts. In this chapter, we look at emerging technologies and attempt to assess their impact on the mainstream of solid-state technology.

13.1 | Dominant Trends in Technology: Substrate Size and Device Density

Two trends in silicon technology have significantly decreased the cost per function of circuits: the use of continually larger wafers, and the ability to manufacture devices with smaller geometries and, hence, higher packing densities. The introduction of larger silicon substrates every few years has served to continually push processing technology forward. To prevent a decrease in yield every time a larger substrate size is introduced, processing technology undergoes improvements to maintain the same uniformity across these larger wafers. Economics is the reason for the use of ever-larger wafers—it costs only slightly more to process each larger wafer, but the total number of good dice with the same yield percentage is proportional to the increase in area. The ability of crystal manufacturers to produce ever-larger wafers seems to be a continuing phenomenon, and one wonders about the size of silicon wafers in the future. The use of ever-larger wafers requires that the thickness of the wafers also be increased to have wafers with the same resistance to breakage.

The early 1980s saw the movement to 100-mm (4-inch) wafers. By 1985, all new fabrication areas were being built to accommodate 150-mm (6-inch) wafers. Also, 200-mm (8-inch) wafers were available, illustrating that substrate manufacturers were ahead of their counterparts in the process equipment industry.

The use of a larger wafer diameter every few years continues to alter the way wafers are handled and processed. The picture of a wafer as a small, light, and fragile piece of silicon is no longer accurate when the wafer is 150 to 200 mm (6 to 8 inches) in diameter. Actions such as physically loading a boat of wafers into a furnace which is five feet high cause problems that are not experienced with lighter wafers. Equipment design must continually evolve to provide satisfactory process uniformity over ever-increasing areas.

13.2 Alignment-and-Exposure Step

The alignment-and-exposure step discussed in the section on photomasking is another area in which significant technological strides are being made. In conventional alignment systems, the wavelength of the light used to expose the photoresist limits the minimum dimensions that can be transferred from a mask to a wafer because of the diffraction of the light. (Diffraction is the bending of light as it passes an edge.) If any space is left between the mask and the wafer during exposure, diffraction occurs. The minimum dimension that could be transferred if everything is optimal is two to four times the wavelength of the light used. However, practical problems, such as wafer flatness, mask flatness, particles, etc., make such dimensions impossible in a production situation. The wavelength of the illumination from the mercury arc lamp is about 4000 Å, so a practical minimum for the line width, using optical techniques, is between 1.0 and 2.0 μm. A survey of contemporary technology shows that some devices are within a factor of two of these dimensions already.

The fundamental limitation is the wavelength of the light used to expose the photoresist, so performing the exposure with a shorter wavelength offers a possible solution to this problem. Two possible alternative methods of exposing a "sensitized" resist layer are to use X-rays or to use electrons.

X-rays generated from a number of sources have low enough energy not to pose a problem to human beings. The wavelengths of these "soft" X-rays

13.2 Alignment-and-Exposure Step

range from 5 to 15 Å. This wavelength is also short enough not to impose any serious diffraction problems. With a source of X-rays and an X-ray sensitive resist, all that is needed is a mask and a method of aligning the mask to the wafer. A diagram of an X-ray exposure system is shown in Figure 13–1.

The generation of a mask for such a system, and the problem of aligning the mask to an image already present on the wafer, are subjects which have been studied. The operation is complicated by the need to expose the wafers in a vacuum because of the limited distance the X-rays travel in air. Devices are already being manufactured in laboratories by means of X-ray exposure techniques. The commercial viability of X-ray exposure in volume production remains to be proven.

The use of electron-beam exposure offers many of the advantages of X-ray exposure, as well as a few others. Electrons are charged particles, but may be viewed as having wave properties. The equivalent wavelength of electrons used for exposure systems is less than an angstrom. Electron beams can be generated using contemporary equipment, as shown in Figure 13–2. An electron beam exposes the resist in the areas where the electrons strike the resist layer.

Thus far, X-ray and electron-beam exposure systems look quite comparable, but there are two advantages to an electron-beam exposure system that have not been mentioned. First, the electron-beam microscope is an instrument used for high magnifications. Modifying the electron-beam system for alignment as well as magnification solves the mask-to-wafer alignment problem. Second, because an electron beam is a stream of charged particles, it is possible to deflect the beam as well as turn it on and off. This capability means that no mask is needed to expose a layer of resist. Instead, the beam can be scanned and turned on and off to directly accomplish the exposure. Thus, the possibility of exposing wafers by using a computer-controlled scanning electron beam is attractive to manufacturers.

Electron-beam equipment is being used commercially to manufacture photomasks. The rapid-scanning feature eliminates the need to step and repeat the pattern, and the entire mask may be exposed during one scan. Electron-

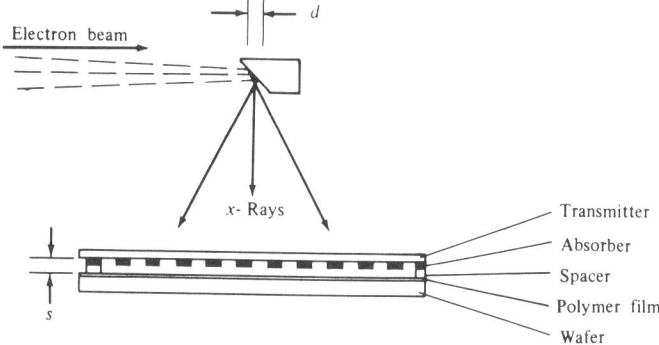

Figure 13–1: Schematic diagram of the soft X-ray lithographic system.

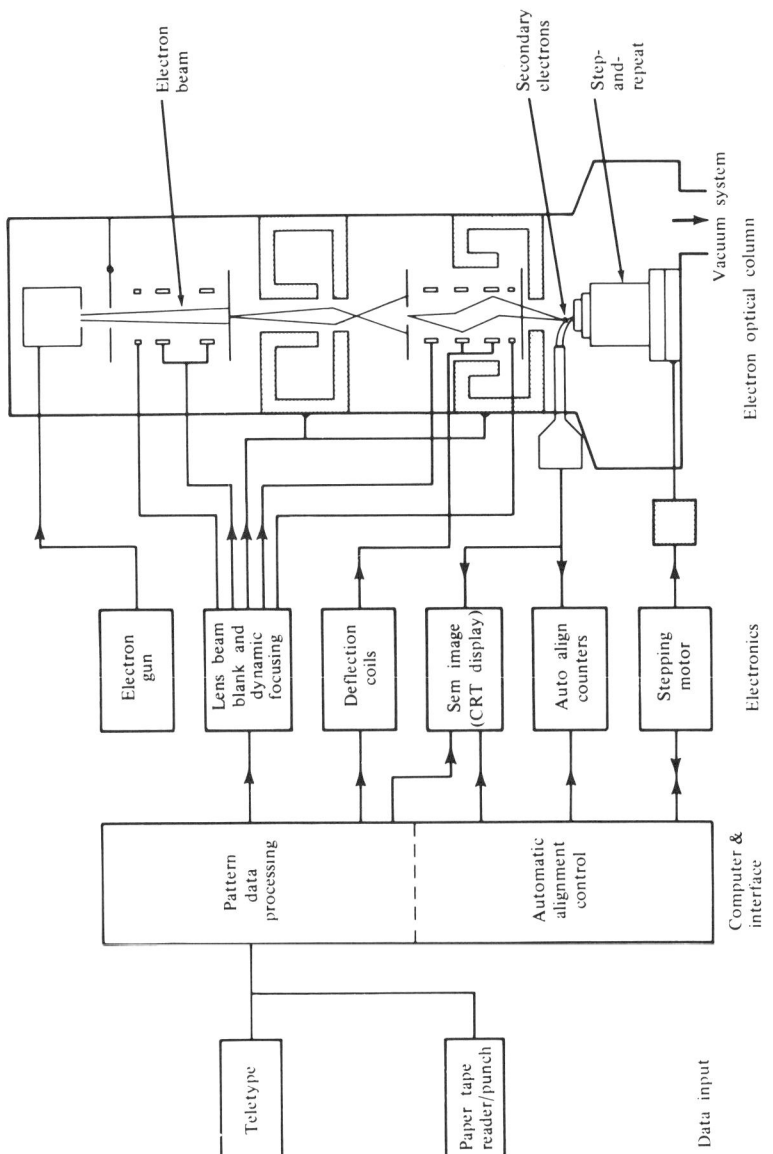

Figure 13–2: Electron–beam exposure system.

beam exposure systems for conventional wafer exposure are not used outside of laboratories and advanced prototype lines because of high cost, low throughput, and the skill needed to operate the equipment.

13.3 Developments in Other Processing Areas

The smaller feature sizes required by new generations of circuits are affecting other processing areas also—for example:

1. *Epitaxial deposition.* In addition to requiring uniformity across larger substrates, thinner epi layers are required. The decreasing surface dimensions are being matched by decreasing vertical dimensions.
2. *Oxidation.* Thinner layers with high dielectric strengths are required, particularly for the gate dielectric. Gate oxides only a few hundred angstroms thick are being used in manufacturing, while much thinner layers are being investigated.
3. *Etching.* Dry-etching techniques for all layers are being used for feature sizes below 3 μm. In many etch systems, wafers are individually etched to guarantee that the needed uniformity is obtained.
4. *Impurity introduction.* Ion implantation is being used almost exclusively on wafers larger than 100 mm (4 inches) to obtain acceptable distributions.
5. *Conductive layers.* Conventional metallization is being supplemented by refractive metals (metals, such as tungsten, that are capable of withstanding high-temperature processing steps), silicides (compounds of silicon and specific metals such as tungsten and molybdenum), and polycides (combinations of polycrystalline silicon and certain metals such as molybdenum and tungsten). Interconnects of these three materials significantly reduce resistance, compared to doped polycrystalline silicon.
6. *Metallization.* Multilayer (two or more layers) metallization allows denser packing of the integrated circuits and, hence, smaller circuits.
7. *Passivation.* Plasma nitride and spin-on polymers such as polyimide are replacing doped SiO_2 as the final passivating layer.

13.4 Developments in Device Technology

Many recent developments in device technology show the possibility of significantly affecting the direction of technology. The first of these developments was the discovery of the *charge-coupled device,* or the CCD. In its simplest form, a CCD consists of a lightly doped silicon substrate with a thin layer of SiO_2 on it, and a series of metallized electrodes on top of the oxide layer, as shown in Figure 13–3. The charge is stored in potential "wells" created by the voltage on the electrodes. To transfer the charge, a deeper potential well is created next to the storage site, causing the carriers to fall into the next well, as shown in Figure 13–4. By varying the potential along a series of electrodes, the "packet" of charge is made to move from one location to another on the semiconductor chip.

The relative simplicity of the CCD structure means that fewer fabrication steps are required in its manufacture. Fewer fabrication steps imply that large-area circuits are economical. Circuits that require large areas include optical

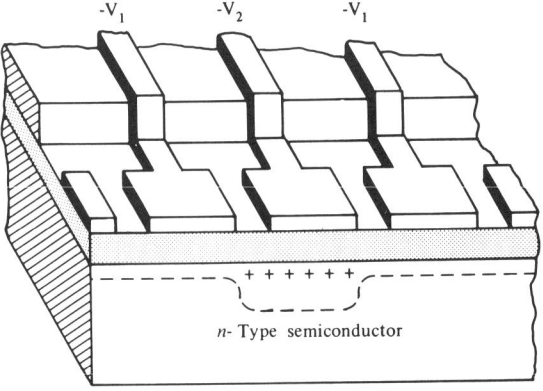

Figure 13–3: A CCD in the storage mode. $-V_2$ is greater than $-V_1$, forming a potential well that captures charge.

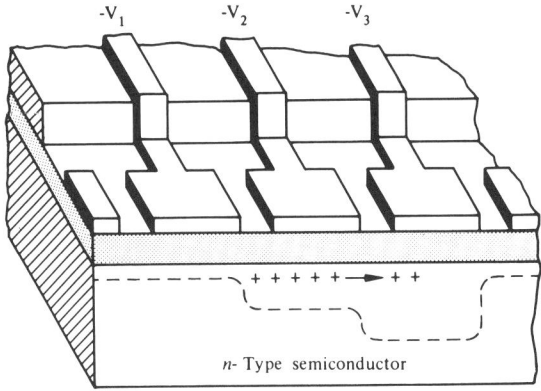

Figure 13–4: Charge is transferred to the next electrode when a still larger voltage $-V_3$ creates an even deeper potential well.

13.4 Developments in Device Technology

sensors and signal processors. Application of CCD technology to these areas is already under way, resulting in a series of successful products. The most visible of these are the small television cameras that are now available. CCD image sensors have replaced large tubes in the camera, significantly reducing the overall size.

Silicon has properties that also allow it to be used to manufacture other types of sensors besides optical sensors. Some of these applications are given in the following list:

1. *Particle detectors.* Doping silicon crystals with nonstandard materials such as lithium and some heavy metals forms regions that are sensitive to the passage of particles such as electrons and neutrons.

2. *Pressure transducers.* If the wafer is etched until it is very thin in regions containing resistors, changes in pressure then cause the thin silicon diaphragm to "flex" change the value of the resistors. This change in resistance is related to the change in pressure and may be used to measure it. The phenomenon is called the *piezoresistive effect.* Figure 13–5 shows the cross section of a pressure transducer manufactured in silicon.

3. *Temperature sensors.* Changes in temperature affect resistors, diodes, and other types of devices in a predictable fashion. By introducing precise amounts of dopants into the silicon to form these devices, both absolute temperatures and changes in temperature may be measured.

4. *Magnetic field sensors.* The presence of a magnetic field affects the flow of carriers in silicon through a phenomenon known as the *Hall effect.* By

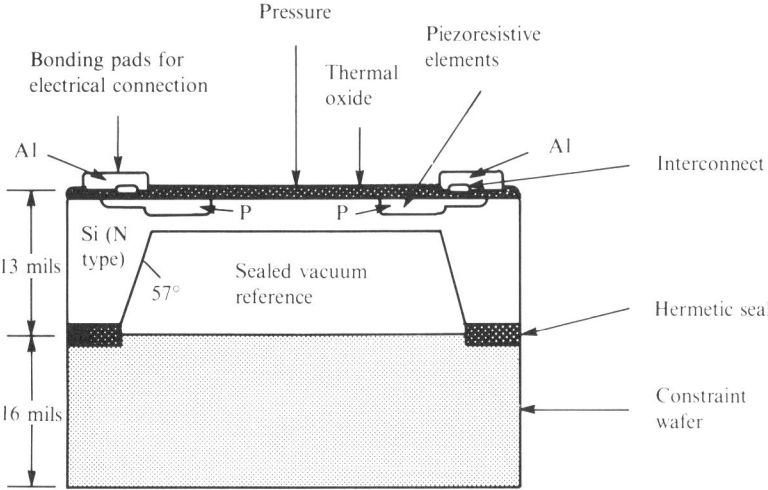

Figure 13–5: Cross section of a pressure transducer in silicon.

manufacturing devices such as resistors and transistors that have been optimized for this application, magnetic fields may be measured.

5. *Light sensors.* Diodes, resistors, transistors, and other devices may be manufactured so that they are sensitive to light. They may then be used to do such things as turn a bulb off during daylight or trip an alarm if a beam of light is interrupted.

The possibility of using silicon for a variety of applications besides the manufacture of ICs considerably broadens its range of use.

Review Exercises

1. Two methods of significantly reducing device sizes involve the use of what types of beams?
2. What is the principal advantage of using larger silicon wafers? What limits the maximum size of a silicon wafer?
3. What is the minimum dimension that could be transferred from mask to wafer using yellow light (0.6-μm wavelength)?
4. What is the minimum practical dimension that can be transferred from mask to wafer using today's semiconductor technology?
5. What are two major advantages of the electron-base exposure system?
6. Sketch and explain the operation of a typical CCD structure.
7. What is a major advantage of CCD technology?
8. List three types of input besides light that silicon can detect.

14 | Nonsilicon technology

14.0 | Introduction

The technology that has been developed to manufacture silicon devices and integrated circuits has other applications. It may be used to manufacture devices in other semiconductors besides silicon. It may also be used to manufacture structures using materials that are not semiconductors. This chapter explores the extension of silicon technology to both of these areas.

14.1 | Light-Emitting Diodes and Laser Diodes

Light-emitting diodes (LEDs) are fabricated in III-V semiconductor compounds such as gallium arsenide and gallium arsenide-phosphide. These semiconductors are called III-V semiconductors because one of the constituents of the compound is from group III of the periodic table, while the other constituent is from group V. The electron configuration of certain of these compounds makes the emission of light possible when current flows through a diode that has been properly fabricated. The technology used to fabricate diodes in these semiconductors closely parallels standard silicon technology. The processing steps are as follows:

1. Starting with the proper substrate, grow a carefully tailored layer of epitaxy.
2. Cover the front surface of the substrate with a protective layer of silicon dioxide using a low-temperature chemical vapor deposition technique.

3. Transfer an image to the front surface of the wafer by using a photolithographic process:
   a. Apply a layer of photoresist.
   b. Using a mask, transfer a pattern to the photoresist.
   c. Using the photoresist as a mask, remove the deposited silicon dioxide in selected regions by etching.
4. Perform a high-temperature diffusion to form the diode.
5. Form ohmic contacts to the diode.
6. Separate the diodes on a substrate into discrete diodes or arrays.
7. Test and package the LEDs.

This general fabrication sequence may be used with different materials to fabricate LEDs that emit red, yellow, green, and even blue light.

If extreme care is taken in the structure and the materials used in the diode, *laser diodes* may be fabricated. (The term *laser* is an acronym for *l*ight *a*mplification by *s*timulated *e*mission of *r*adiation.) A laser diode emits a light with just a single wavelength. Laser diodes are useful for generating the light transmitted using fiber optics.

14.2 Optical Integrated Circuits

One eventual use of light-emitting diodes or laser diodes may be in optical integrated circuits. At the same time that light-emitting regions are fabricated in a substrate, it is possible to fabricate light-sensitive regions. Recent work indicates that it may be possible to link a light-emitting and a light-sensitive region with an intermediate region that serves as a light "wave-guide." With the successful fabrication of light generation, light transmission, and light reception regions on the same substrate, the fabrication of an optical integrated circuit will be possible. It will then remain for engineers and scientists to prove the usefulness of such circuits in specific applications.

14.3 Liquid Crystal Displays

A rival to LEDs in many applications, particularly in the display market for portable instruments, is the *liquid crystal display*, or LCD. The name of the devices describes its composition: LCDs have a liquid crystal material sandwiched between two glass plates separated by about 12 μm. The glass plates are sealed around their perimeter. The image appears between two layers of conductive material, one etched on the surface of each glass plate.

In operating the device, a voltage is applied between the two layers of conductive material. This voltage changes the molecular orientation of the liquid crystal material. The light passing through the display is thus scattered or reflected unequally where the voltage has been applied. LCDs emit no light, but can use either reflected or transmitted light to form the display. They have the large advantage of consuming considerably less current than LEDs. For this reason, many portable instruments, such as watches, calculators, and portable computers, use LCDs.

14.4 Gallium Arsenide Transistors and ICs

Gallium arsenide (GaAs) has properties that make it attractive for applications requiring high-speed transistors or circuits. The mobility of carriers (the speed with which the carriers move when a voltage is applied) in GaAs is two to three times greater than the mobility of carriers in silicon. When transistors with the same dimensions are fabricated in GaAs, they are faster by the same ratio. For this reason, GaAs transistors are used at frequencies where silicon transistors do not have acceptable performance.

Developing the techniques for fabricating integrated circuits out of GaAs has taken many years because the processing technology is more difficult than fabricating silicon-based circuits. The cross section of a GaAs field effect transistor (called a *depletion mode MOSFET* because of its operation) is shown in Figure 14–1. GaAs does not grow a protective layer on its surface at high temperatures; instead, it decomposes. (Silicon grows SiO_2, which is the key to silicon processing technology.) Processing techniques suitable for GaAs IC manufacture have taken longer to develop, but logic and memory ICs are commercially available. The cost of these circuits makes them attractive only in applications needing the highest speeds, such as ultrafast instruments and computers.

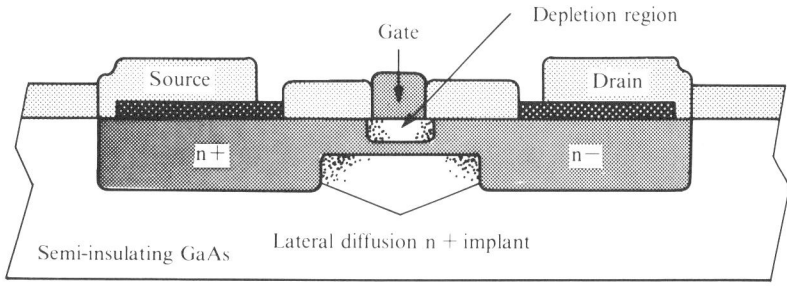

Figure 14–1: Cross section of a gallium arsenide field-effect transistor.

14.5 Josephson Junctions

At temperatures very close to absolute zero ($-273°C$), the resistance of certain materials drops to zero. These materials, known as *superconductors*, may be used to form a device known as a *Josephson junction*. This device is made by separating two superconducting regions using a thin layer of insulating material. The conduction properties at these temperatures produce a "switching" phenomenon that may be used for digital circuits. The main advantages of Josephson junctions are high speed and low power dissipation. However, problems associated with both device fabrication and operation at the low temperatures required most be solved if this technology is to be commercially successful.

14.6 Quartz Crystal Oscillators

Silicon dioxide in its crystalline form (quartz) is a piezoelectric material. When it is placed in an electric circuit with a voltage across it, it vibrates, producing an output whose frequency is a function of the physical parameters of the crystal. Since quartz is silicon dioxide, it is not surprising to find the microelectronic technology used in the manufacture of timing devices from quartz. The specific steps used in their manufacture are centered around the transfer of a metal pattern to both sides of a quartz wafer. The ability to produce large numbers of quartz timers simultaneously is the main reason for the use of microelectronics technology. Quartz oscillators are the timing element used in most analog and digital watches. The stability of the frequency generated using a crystal oscillator is better than that of the mechanical watch movements they have replaced.

14.7 Magnetic Bubble (Magnetic Domain) Devices

A phenomenon in which magnetic "bubbles" are formed and made to move and interact in predetermined ways is the basis of another type of device. The bubbles are actually cylindrical magnetic domains whose polarization is opposite to that of the thin magnetic film in which they are embedded. The bubbles are stable over a wide range of conditions and can be moved from one point to another at high velocity. The technologies used to form the control regions on the surfaces of the thin magnetic film are similar to those used in the fabrication of integrated circuits. Memories made using magnetic bubbles rival those using standard semiconductors in some applications. Bubble memories retain their information if the power is removed and are not altered by a noisy electrical environment. These characteristics make them attractive in certain instances where conventional memories do not perform well.

14.8 | Hybrid Technology

Hybrid technology is the technology of placing devices and circuits on a common substrate and interconnecting them to perform useful electrical functions. Hybrid technology is divided into two fabrication technologies: thin film and thick film.

Thin-film hybrid circuits are manufactured by first depositing a thin layer of metal on a substrate using vacuum deposition techniques. Next, the substrates are coated with photoresist, baked, and exposed, and the metal layer is selectively removed by etching to form the desired pattern. Active and passive devices such as diodes, resistors, transistors, and capacitors are attached to the substrate to complete the circuit.

Thick-film structures are prepared by screening and firing or by pyrolytic deposition. They generally contain only conductors, resistors, and capacitors, with the other components added as discrete entities. All are put down on a substrate which is usually alumina. A thick film is a conductive, resistive, or insulating film thicker than 10 mils that is produced by firing a paste deposited on a substrate. The paste is deposited on the substrate by the stencil screen printing process. Different paste compositions are used for various components in the circuit. After the pattern has been screened onto a ceramic substrate and dried, it is fired in a furnace, where the composition of the paste gives rise to the final characteristics of each layer.

Although thick-film technology is usually less costly than thin-film technology, it requires a larger substrate to accommodate the same circuit complexity. Generally speaking, a thick-film circuit is limited in resistor tolerances to no better than $\pm 1\%$, and in resistances to less than 5 megohms. Thick-film techniques are generally used to build circuits that operate below 1 gigahertz and that do not require the tolerances and line precision obtainable with thin-film techniques. Thin-film fabrication, on the other hand, is ideal for high-frequency microwave applications and those requiring highly precise line widths and circuit elements.

Hybrid technology allows devices to be used together that cannot be simultaneously fabricated. By selecting the characteristics of each device type, performance may be achieved in hybrid circuits that is beyond that of monolithic (one-chip) integrated circuits.

Review Exercises

1. Does an LED or an LCD display require more power?
2. What effect is responsible for the use of quartz as a timer?
3. List and describe the processing steps required to produce light-emitting diodes.
4. How do a laser diode and an LED differ?
5. Why are GaAs transistors faster than silicon transistors?
6. Why has it taken so long to develop GaAs ICs?
7. Where might bubble memories be used instead of semiconductor memories?
8. What is the basic difference between thin-film and thick-film hybrid circuits?

APPENDIX 1
Scientific notation

A1.1 | Mathematics

In dealing with physically measurable amounts or numbers of materials, very large and very small numbers are usually encountered. The writing of these numbers usually means using many zeros. To avoid the continuous writing of zero, mathematicians devised a shorthand method usually referred to as scientific notation. The idea behind scientific notation is built around the number 10 and can be developed as follows. The number 10 can also be written as 10 to the first power, or 10^1. The number 1 is the power to which 10 is raised. (The power to which a number is raised is the number of times that number is multiplied by itself.) If we multiply 10 by itself, we get $10 \times 10 = 100$. But this is two 10s multiplied together, so $10 \times 10 = 10^2$ (10 to the second power) $= 100$. In a similar fashion, $10 \times 10 \times 10 = 10^3 = 1000$. A power of 10 greater than zero tells us the number of zeros to the right of the 1 in the number we are dealing with. Following are some common multiples of 10 with powers greater than zero.

$$10 = 10^1 = 10 \text{ to the first power}$$
$$100 = 10^2 = 10 \text{ to the second power}$$
$$1000 = 10^3 = 10 \text{ to the third power}$$
$$10{,}000 = 10^4 = 10 \text{ to the fourth power}$$
$$100{,}000 = 10^5 = 10 \text{ to the fifth power}$$
$$1{,}000{,}000 = 10^6 = 10 \text{ to the sixth power}$$

Numbers smaller than one can also be represented using scientific notation. The number $1/10 = .1$ is written as 10^{-1} (10 to the minus one power). Similarly, the number $1/100 = 1/10^2 = .01$ is written as 10^{-2} (10 to the minus two power). A power of 10 less than zero tells the number of places the decimal point is to the left of the 1 in the number we are dealing with. More common multiples of 10 with powers less than zero are listed here.

$$.000001 = 10^{-6} = 10 \text{ to the minus sixth power}$$
$$.00001 = 10^{-5} = 10 \text{ to the minus fifth power}$$
$$.0001 = 10^{-4} = 10 \text{ to the minus fourth power}$$
$$.001 = 10^{-3} = 10 \text{ to the minus third power}$$
$$.01 = 10^{-2} = 10 \text{ to the minus second power}$$
$$.1 = 10^{-1} = 10 \text{ to the minus first power}$$

In most cases, it is convenient to rewrite large or small numbers as numbers between one and 10 times 10 to some power. To express a number greater than 10 in this form, move the decimal one place to the left and count the number of places from the original decimal point. The number of places counted will give the correct positive power of 10.

EXAMPLES:

$$942 = 9.42 \times 10^2$$
$$420{,}610 = 4.2061 \times 10^5$$
$$31 = 3.1 \times 10^1$$

To express a number less than 1 in this form, move the decimal point to the right until it is part of the first nonzero number. Count the number of places the decimal point has been moved from its original position. The number of places counted is the negative power of 10.

EXAMPLES:

$$.00882 = 8.82 \times 10^{-3}$$
$$.0000031 = 3.1 \times 10^{-6}$$
$$.064 = 6.4 \times 10^{-2}$$

The only power of 10 we have not discussed so far is 10^0. The value of 10^0

is 1. This statement makes sense when considering the values of 10^{-1} and 10^1.

$$10^{-1} = .1$$
$$10^0 = 1$$
$$10^1 = 10$$

A1.2 Addition and Subtraction

Before numbers written using scientific notation can be added or subtracted, all numbers must be written using the same power of 10. Once the power of 10 of the numbers is the same, addition and subtraction is performed, and the answer is the resulting number to the common power of 10.

EXAMPLE 1:

$$4.13 \times 10^3$$
$$+9.64 \times 10^5$$

First, rewrite 9.64×10^5 as 964×10^3. Then rewrite the problem.

$$\begin{array}{r} 4.13 \times 10^3 \\ +964.00 \times 10^3 \\ \hline 968.13 \times 10^3 = 968{,}130 \end{array}$$

The solution can be checked, since

$$4.13 \times 10^3 = 4130 \text{ and } 9.64 \times 10^5 = 964{,}000$$

This sum is just

$$\begin{array}{r} 964000 \\ +4130 \\ \hline 968130 \end{array}$$

EXAMPLE 2:

$$4.93 \times 10^7$$
$$-9.4 \times 10^5$$

At first glance, it appears that subtraction will result in a number less than zero. But the first number is rewritten as

$$4.93 \times 10^7 = 493 \times 10^5$$

The subtraction is then done, keeping the common power.

$$\begin{array}{r} 493 \times 10^5 \\ - 9.4 \times 10^5 \\ \hline 483.6 \times 10^5 \end{array}$$

A1.3 | Multiplication

Multiplication of numbers written in scientific notation is accomplished using the following rules:

1. Multiply the number to the left of the power of 10 to obtain the numerical part of the product.
2. Add the powers of 10 together to get the resultant power of 10.

EXAMPLE 1:

$$(4.3 \times 10^4) \times (7.6 \times 10^2)$$

$$\left. \begin{array}{l} 4.3 \times 7.6 = 32.68 \\ 10^4 \times 10^2 = 10^6 \end{array} \right\} = 32.68 \times 10^6 = 3.268 \times 10^7$$

EXAMPLE 2:

$$(1.2 \times 10^{-4}) \times (9.1 \times 10^6)$$

$$\left. \begin{array}{l} 1.2 \times 9.1 = 10.92 \\ 10^{-4} \times 10^6 = 10^2 \end{array} \right\} = 10.92 \times 10^2 = 1.092 \times 10^3$$

A1.4 | Division

The rules for division of numbers written using scientific notation are given below:

1. Divide the numbers to the left of the power of 10 to obtain the numerical portion of the answer.

A1.4 Division

2. Subtract the power of 10 of the denominator from the power of 10 of the numerator to obtain the power of 10 of the answer.

EXAMPLE 1:

$$\frac{9.81 \times 10^7}{4.19 \times 10^3}$$

$$\left.\begin{array}{l} \dfrac{9.81}{4.14} = 2.34 \\[6pt] \dfrac{10^7}{10^3} = 10^{7-3} = 10^4 \end{array}\right\} = 2.34 \times 10^4$$

EXAMPLE 2:

$$\frac{3.1 \times 10^{-3}}{5.6 \times 10^3}$$

$$\left.\begin{array}{l} \dfrac{3.1}{5.6} = .554 \\[6pt] \dfrac{10^{-3}}{10^3} = 10^{-3-(3)} = 10^{-6} \end{array}\right\} = .554 \times 10^{-6} = 5.54 \times 10^{-7}$$

APPENDIX 2
Use of graphs

Throughout the text, much of the information presented is in the form of graphs. Graphs are a way of presenting much information in a small space, without getting into the complicated mathematics often involved. A typical graph is shown in Figure A2–1.

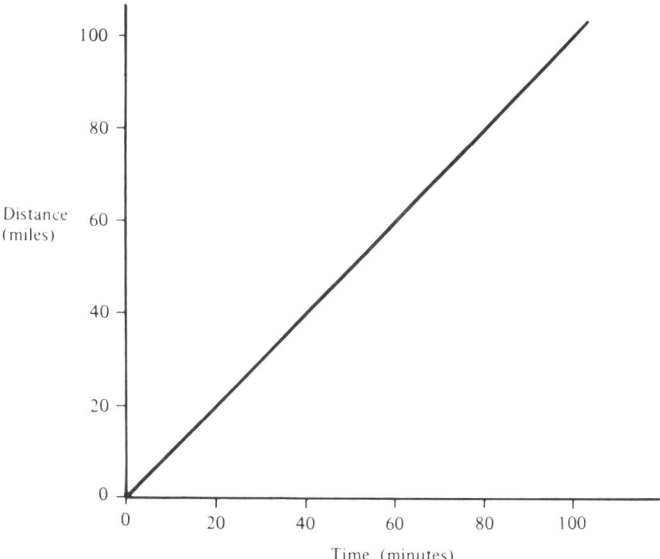

Figure A2–1: Graph of the distance traveled by a car versus time for a constant velocity.

This graph has time in minutes along the horizontal axis and distance in miles along the vertical axis. If a car is traveling at 60 miles per hour (or 1 mile per minute), the line represents the distance traveled for any elapsed time. To determine the distance covered in 30 minutes, find 30 minutes on the horizontal axis, and follow a path straight upward until it intersects the line. Then follow a path straight across to the vertical axis. The point of inter-

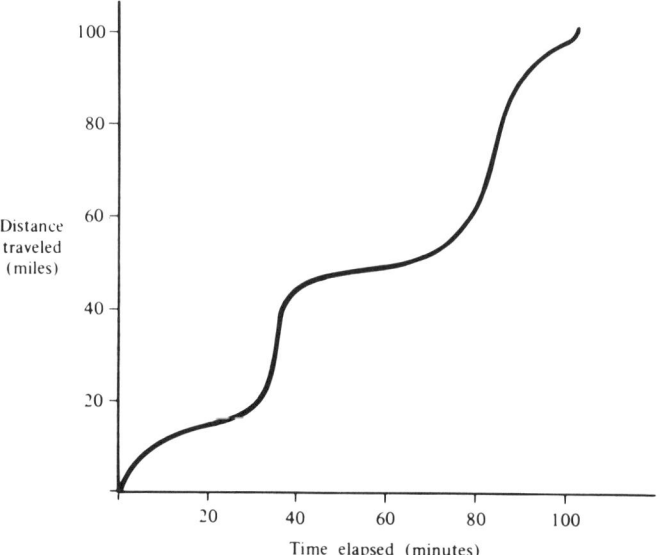

Figure A2-2: Graph of the distance traveled by a car versus time for a varying velocity.

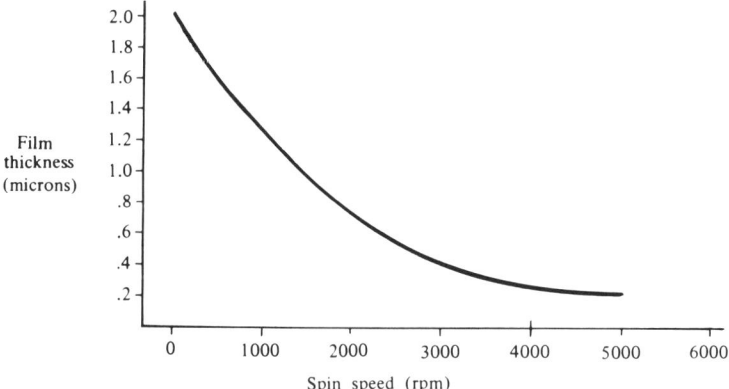

Figure A2-3: Film thickness as a function of application spin speed.

Appendix 2: Use of Graphs

section should be at 30 miles along this axis. In a similar fashion, the distance traveled for any time up to 100 minutes can be found using the graph. The information provided by Figure A2–1 can be easily obtained by multiplying the time elapsed by one mile per minute, to obtain the total distance traveled. But if a constant speed is not used, a graph such as that shown in Figure

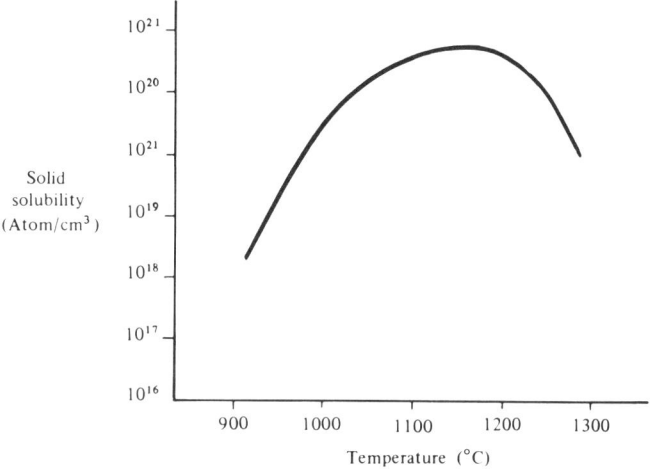

Figure A2–4: Solid solubility of elements in silicon as a function of temperature.

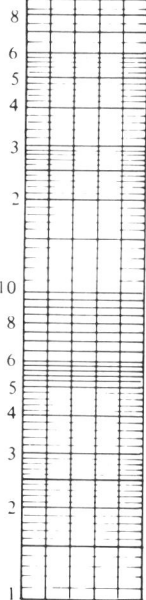

Figure A2–5: Extended section of the logarithmic scale.

A2–2 would result. At the end of 100 minutes, 100 miles have been covered, but at times between zero and 100 minutes, a set relationship between the time elapsed and the distance traveled is not easily obtained. This graph provides a method of determining the relationship between the time elapsed and the distance traveled without requiring the use of complicated mathematics.

There are three types of graphs that we will be concerned with. Each has a different type of coordinate scale along each axis. The first type of graph has the variables displayed in a linear fashion along each axis, as in Figures A2–1 and A2–2. Another graph of this type is shown in Figure A2–3, which shows the film thickness that results when a material is applied to a wafer using a

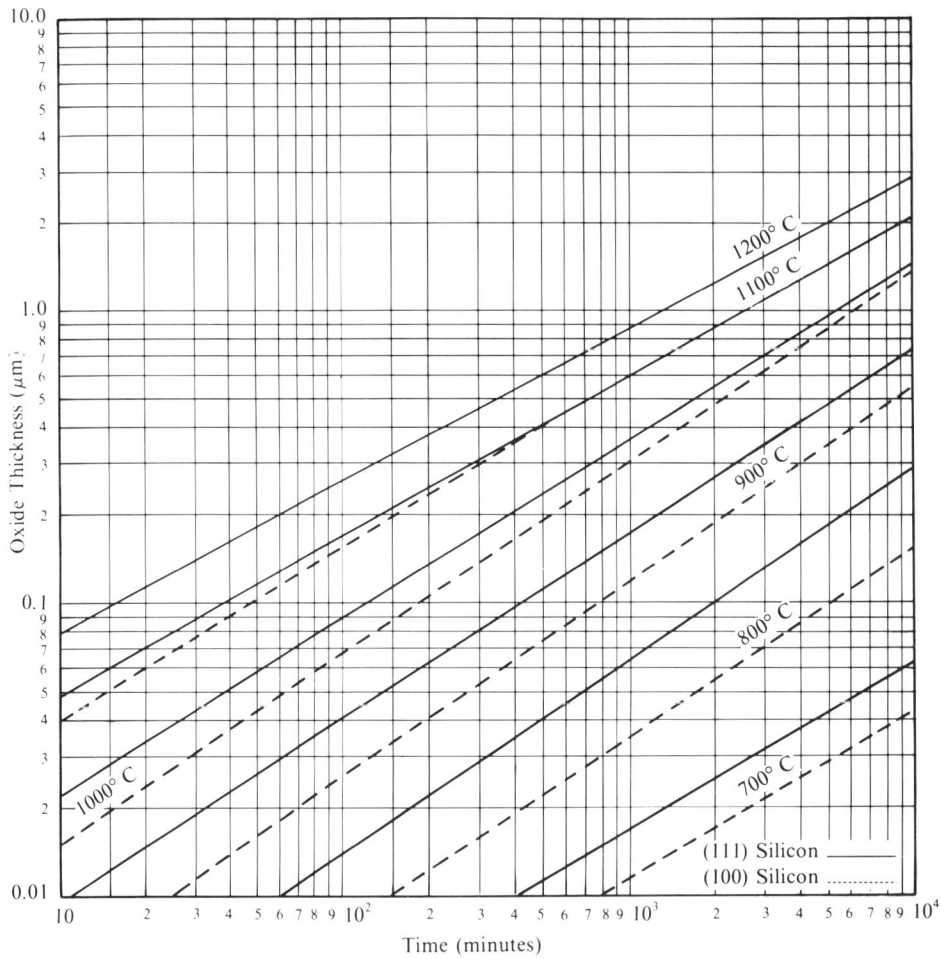

Figure A2–6: Oxide thickness versus oxidation time for silicon in dry O_2 [1]. [1] O.D. Trapp et al., *Semiconductor Technology Handbook* Technology Associates (1985), pp. 4.1–4.2.

spinning technique. Both axes have linearly increasing variables, usually starting with zero.

The second type of graph has the variable along one axis displayed using a logarithmic scale, as shown in Figure A2–4. The horizontal scale is linear, but the vertical scale increases by a factor of 10 between every major line. Use of a logarithmic scale allows information with a large numerical range to be displayed. An expanded portion of the logarithmic side is shown in Figure A2–5. In Figure A2–4, the numbers between 10^{17} and 10^{18} are numbered 2,3, etc. The line at 2 corresponds to 2×10^{17}. The next line corresponds to 3×10^{17}, and so on, until we reach 9×10^{17}. The line above 9×10^{17} is 10×10^{17} which equals 1×10^{18}. It is marked on the graph. In a similar fashion, the line marked 2 above the 1×10^{18} line is 2×10^{18}. This type of graph is often called a semilog graph.

Information relating two variables that have wide ranges can be displayed using a logarithmic scale along each axis. Such a graph is called a log–log graph, and an example of one is shown in Figure A2–6. (This graph is also Figure 3–9.)

APPENDIX 3
Units

The semiconductor industry is presently undergoing a transition from the use of English units (inches, feet, mils) to the use of metric units (millimeters, angstroms, centimeters). The set of units encountered is a function of the area of the company a person works in. In many semiconductor fabrication areas, surface measurements such as die size and device dimensions are measured in mils, while junction depths and epitaxial layer thicknesses are measured in angstroms or micrometers. The ability to rapidly convert from one system of units to another is a necessity. Table A3–1 is designed to assist in the conversion from one set of units to another. To accomplish conversion, find the first set of units on the left side and the second set of units along the top. The number at the intersection of the row and column is the conversion factor.

EXAMPLE 1:

Convert 2.3 mils to micrometers.

SOLUTION:

Following the rule just given, we find that the number at the intersection of the "mil" row and the "micron" column is 25.4. So

$$2.3 \text{ mils} \times 25.4 \frac{\text{micrometers}}{\text{mil}} = 58.4 \text{ micrometers}$$

Table A3–1: Length Units Used in Semiconductor Technology

TO GET → FROM ↓	MULTIPLY BY → INCH (″)	MIL	MICROINCH	CENTIMETER (cm)	MILLIMETER (mm)	MICROMETER (μm)	NANOMETER (nm)	ANGSTROM (Å)
Inch (″)	1	10^3	10^6	2.54	25.4	2.54×10^4	2.54×10^7	2.54×10^8
Mil	10^{-3}	1	10^3	2.54×10^{-3}	2.54×10^{-2}	25.4	2.54×10^4	2.54×10^5
Microinch	10^{-6}	10^{-3}	1	2.54×10^{-6}	2.54×10^{-5}	2.54×10^{-2}	25.4	2.54×10^2
Centimeter (cm)	0.3937	3.937×10^2	3.937×10^5	1	10	10^4	10^7	10^8
Millimeter (mm)	3.937×10^{-2}	39.37	3.937×10^4	0.1	1	10^3	10^6	10^7
Micrometer (μm)	3.937×10^{-5}	3.937×10^{-2}	39.37	10^{-4}	10^{-3}	1	10^3	10^4
Nanometer (nm)	3.937×10^{-8}	3.937×10^{-5}	3.937×10^{-2}	10^{-7}	10^{-6}	10^{-3}	1	10
Angstrom (Å)	3.937×10^{-9}	3.937×10^{-6}	3.937×10^{-3}	10^{-8}	10^{-7}	10^{-4}	10^{-1}	1

Appendix 3: Units

EXAMPLE 2:

Convert 32 microinches to angstroms.

SOLUTION:

The number at the intersection of the microinch row and the angstrom column is 2.54×10^2. So

$$(32)(2.54 \times 10^2) = 8128 \text{ Å} = .8128 \text{ micrometers}$$

APPENDIX 4
Solutions

A4.1 Chemistry and Physics of Semiconductor Materials

1. Since only phosphorus atoms have been added and they are donors,
   a. $N_D = 10^{15}/cm^3$
   b. $N_A = 0$
   c. All of the donors ionize, with each donor producing one conduction electron. Therefore,
   $$n = 10^{15}/cm^3$$
   d. We can solve for p, since $n \cdot p = n_i^2$ and $n_i^2 = 2 \times 10^{20}/cm^6$
   $$p = \frac{n_i^2}{n} = \frac{2 \times 10^{20}}{10^{15}} cm^3 = 2 \times 10^5/cm^3$$

   Figure 1–7 can be used to determine the resistivity.
   $$\rho = 5 \; \Omega\text{-cm}$$

2. Since only boron has been added to the silicon, and it is an acceptor atom,
   a. $N_D = 0$
   b. $N_A = 2 \times 10^{16}/cm^3$

c. Every acceptor atom produces one hole, so $p = N_A = 2 \times 10^{16}/cm^3$. We can solve for n since $n \cdot p = n_i^2$.

$$n = \frac{n_i^2}{p} = \frac{2 \times 10^{20}/cm^6}{2 \times 10^{16}/cm^3} = 1 \times 10^4/cm^3$$

d. $p = 2 \times 10^{16}/cm^3$

e. Figure 1–7 can be used to determine the resistivity.

$$\rho = 1 \; \Omega\text{-cm}$$

3. a. Since arsenic atoms are donors, $N_D = 3 \times 10^{17}$ atoms/cm³.
   b. Likewise, boron atoms are acceptors, so $N_A = 5 \times 10^{17}$ atoms/cm³.
   c. Since donors and acceptors cancel each other out, 3×10^{17} atoms/cm³ of the acceptor atoms cancel the effect of 3×10^{17} donor atoms. This leaves only 2×10^{17} acceptor atoms free to give holes. Therefore, $p = N_A - N_D = 2 \times 10^{17}/cm^3$. Since

$$n \cdot p = n_i^2, \; n = \frac{n_i^2}{p} = \frac{2 \times 10^{20}/cm^6}{2 \times 10^{17}/cm^3} = 1 \times 10^3/cm^3$$

   d. $p = 2 \times 10^{17}/cm^3$

4. From the text,

$$R_S = 4.5 \frac{V}{I} = (4.5) \frac{5 \times 10^{-3} \text{ volts}}{4.5 \times 10^{-3} \text{ amps}}$$

$$R_S = 5 \frac{\text{volts}}{\text{amps}} = 5 \text{ ohms}$$

5. The equation for the resistance of a bar of material is

$$R = \frac{\rho L}{\text{width} \times \text{thickness}}$$

$$= (2 \; \Omega\text{-cm}) \frac{100 \text{ micrometers}}{(5 \text{ micrometers})(2 \text{ micrometers})}$$

$$R = \frac{2 \; \Omega\text{-cm}}{\text{micrometers}} \left(\frac{100}{10}\right) = 20 \; \Omega\text{-cm/micrometers}$$

$$R = 200{,}000 \text{ ohms}$$

6. For the conduction electron concentration to equal the hole concentration, the donor and the acceptor concentrations must be equal, or $N_D = N_A$. Presently,

$$N_D = 5 \times 10^{15}/cm^3 \text{ and } N_A = 2 \times 10^{16}/cm^3$$

Addition of 1.5×10^{16} donor atoms/cm^3 will bring the two concentrations into equilibrium.

7. From the dimensions of the bar and its resistance, the resistivity can be determined from

$$R = \frac{\rho L}{W \cdot T}$$

or

$$\rho = \frac{R \cdot T \cdot W}{L} = \frac{(10\Omega)(0.1 \text{ cm})(0.1 \text{ cm})}{1 \text{ cm}} = 0.1 \text{ }\Omega\text{-cm}$$

Since the bar is *n*-type, the donor concentration is found from Figure 1–7 to be

$$N_D = 8 \times 10^{16}/cm^3$$

8. Since the bar contains only acceptor atoms, $n = N_D = 2 \times 10^{15}/cm^3$. The hole concentration is determined using

$$p = \frac{n_i^2}{n} = \frac{2 \times 10^{20}/cm^6}{2 \times 10^{15}/cm^3} = 1 \times 10^5/cm^3$$

The formula for conductivity is

$$\sigma = q(\mu_n n + \mu_p p)$$

But since $n \gg p$, $\sigma \cong q\mu_n n$ and

$$\rho = \frac{1}{q\mu_n n} = \frac{1}{(1.6 \times 10^{-19} \text{ coulombs}) \mu_n (2 \times 10^{15}/cm^3)}$$

From Figure 1–8, $\mu_n = 1400\frac{cm^2}{V\text{-sec}}$. Therefore,

$$\rho = \cfrac{1}{(1.6 \times 10^{-19})(1400)\, 2 \times 10^{15} \cfrac{1}{\text{volt}} \cfrac{(\text{coulomb})}{\text{sec}} \cfrac{\text{cm}^2}{\text{cm}^3}}$$

$$= \frac{1}{(1.6)(1.4)(2) \times 10^{-1}} = 2.2\ \Omega\text{-cm}$$

From Figure 1–7, we have $\rho = 2.2\ \Omega$-cm, so the comparison is quite good.

9. a. In the silicon bar, $N_D = 3 \times 10^{18}$ and $N_A = 1 \times 10^{18}/\text{cm}^3$. Therefore,

   b.
   $$n = N_D - N_A = 2 \times 10^{18}/\text{cm}^3$$
   $$p = n_i^2/n = \frac{2 \times 10^{20}}{2 \times 10^{18}} = 1 \times 10^2/\text{cm}^3$$

   c. The total impurity concentration $C_T = N_A + N_D = 4 \times 10^{18}/\text{cm}^3$, and μ_n and μ_p can be determined from Figure 1–8 using this value for C_T.

   $$\mu_n \cong 170\ \text{cm}^2/V\text{-sec}$$
   $$\mu_p \cong 70\ \text{cm}^2/V\text{-sec}$$

   d.
   $$\sigma = q(\mu_n n + \mu_p p) \cong q\mu_n n \text{ since } n \gg p$$
   $$\rho = \frac{1}{\sigma} = \frac{1}{q\mu_n n}$$
   $$= \frac{1}{(1.6 \times 10^{-19})(170)\, 2 \times 10^{18}}\ \Omega\text{-cm} = 0.018\ \Omega\ \text{cm}$$

   e. This answer differs from that of Figure 1–7 because of the 2×10^{18} atoms that have cancelled each other out, but still modify the mobility of the carriers.

10. The net impurity concentration is found from

    $$N_A - N_D = 7 \times 10^{15}/\text{cm}^3 - 3 \times 10^{15}/\text{cm}^3 = 4 \times 10^{15}/\text{cm}^3$$

    and the material is p-type. Thus, the hole concentration is

    $$p = (N_A - N_D) = 4 \times 10^{15}/\text{cm}^3$$

and the electron concentration is determined by

$$n = \frac{n_i^2}{p} = \frac{2 \times 10^{20}/\text{cm}^6}{4 \times 10^{15}/\text{cm}^3} = 5 \times 10^4/\text{cm}^3$$

A4.2 Crystal Growth and Wafer Preparation

1. It is easier to separate a gas from a liquid or solid than it is to separate just liquids or gases.
2. a. Silicon dioxide
   b. The crucible is made of SiO_2, and some of it melts during the crystal growth process.
3. Silicon dioxide. The silicon dioxide crucible is capable of containing molten silicon, so it must not be molten.
4. a. The crystal orientation determines the preferred direction of wafer breakage.
   b. The break planes of the wafer are denoted by the flat ground on one edge of the wafer.
5. Polysilicon is silicon in which the atoms are not in an ordered crystal structure.
6. The inert atmosphere of argon gas during crystal growth prevents oxidation of the silicon.
7. A seed crystal is necessary to initiate the growth of the ingot with the correct crystal orientation.
8. The two variables which control the diameter of the silicon rod are pull-rate and temperature.
9. The six advantages gained from the edge-rounding step used during wafer manufacture are: removal of microcrack laden corners, reduction of silicon chip contamination, increased ease of loading in processing equipment, reduction of photoresist edge build-up, and improved contact and proximity mask life.
10. Eight quality-control problems that can arise after the sawing operation are: chips, edge flakes, exit damage, bow, taper, non-flatness, microcracks, and silicon-chip contamination.
11. a.
$$x \text{ index} = \frac{1}{1/2} = 2$$

$$y \text{ index} = \frac{1}{1} = 1$$

$$z \text{ index} = \frac{1}{\infty} = 0$$

    b.   We have a ⟨210⟩ plane.

12. A value of $K = 1$ means that the concentration of dopant in the solid equals the concentration of dopant in the liquid.

13. The dopant with K closest to 1 will produce the flattest impurity profile. From the table, this impurity is boron.

14. The two most common orientations are (111) and (100).

15. Slip and dislocation.

A4.3 | Oxidation of Silicon

1. a. Using Figure 3–9, the oxide thickness is 0.2 μm or 2000 Å.
   b. Using Figure 3–10, the oxide thickness is 0.15 μm or 1500 Å.

2. No. The oxidizing species must diffuse through a layer of SiO_2 before it can react. If the oxide growth curve is in the transport-limited region, doubling the time does not result in a doubling of the thickness of the SiO_2 layer.

3. a. Since we are beginning with a bare silicon wafer, we can use Figure 3–9 directly. We find a value of 0.3 μm or 3000 Å.
   b. Using 3000 Å as a starting point on Figure 3–10, we find that we have grown the equivalent of 9 minutes (at 1200°C in steam). We add another 6 minutes, bring the total time to 15 minutes. Figure 3–10 shows a total oxide thickness of 0.4 μm or 4000 Å.
   c. Using 4000 Å as a starting point on Figure 3–10, we find that we have grown the equivalent of 24 minutes at 1100°C steam. Adding another 12 minutes brings this total up to 36 minutes. Figure 3–10 shows a total oxide thickness of 0.5 μm or 5000 Å.

4. In anodic oxidation, silicon is the mobile species.

5. For <100> silicon at 1100°C steam for 24 minutes, an oxide layer 0.4 μm or 4000 Å thick will be grown. In dry O_2 at 1000°C, it takes 1550 minutes to grow 4000 Å and 1800 minutes to grow 4500 Å. The final oxide thickness is 4500 Å and the additional time at 1000°C in dry O_2 is 250 minutes (approximately 4 hours).

6. Water vapor (steam) has a much higher solubility in silicon dioxide than oxygen has. For the same temperature, much more steam will be available in the oxide layer to diffuse to the interface and react.

7. Transport limited implies that the number of available molecules has been limited; whereas, reaction-rate limited implies that the temperature is the limiting factor.
8. Boron tends to be depleted from the silicon during oxidation due to its greater solubility in silicon dioxide.

A4.4 | Photolithography

1. Yes. Positive resist is "light-softened" resist.
2. a. 5500 rpm
   b. 0.8 μm or 8000 Å
3. More viscous.
4. a. Convection heating—hot air is circulated through the chamber heating wafers and carrying away vapor.
   b. IR—infrared radiation heats the wafers evaporating excess solvents.
   c. Conductive heating—the wafer sits directly on a heated plate or chuck, heating the wafer and evaporating the solvent.
5. Iron oxide is transparent to yellow light while being opaque to intense ultraviolet. Chrome masks are very hard and resist scratching while emulsion masks minimize light reflection within the opaque regions.
6. Photolithography is the transfer of an image from the mask to a wafer through the use of photosensitive resist.
7. Photoresist performance is characterized by
   a. Adhesion—a measure of the lateral etch at the edge of a post-baked resist image.
   b. Etch resistance—an oxidized wafer fully coated with photoresist is subjected to an etch several times longer than normal. Any breakdown in the photoresist is looked for.
   c. Resolution—the minimum width and spacing that can be successfully transferred to the resist layer is measured.
   d. Photosensitivity—the absolute response to different light intensities is measured.
8. Priming improves the adhesion of photoresist to certain materials.
9. Spinning.
10. Time and temperature during baking.
11. Develop check verified the quality and alignment of the photoresist pattern.
12. a. Soft bake—excess solvent is evaporated from the resist layer.

b. Hard bake—residual solvent is evaporated from the resist and the adhesion at the edges of the photoresist pattern is increased.

A4.5 Impurity Introduction and Redistribution

1. a. From Figure 5–2, we see that the solid solubility is approximately 2×10^{19} atoms/cm^3.
   b. From Figure 5–2, we see that the solid solubility is approximately 1.1×10^{17} atoms/cm^3.
2. Figure 5–9 reveals that gallium has a higher diffusion coefficient.
3. Using Figure 5–9, we see that this value is 0.71 μm/hr$^{1/2}$.
4. The acceleration energy (in keV or thousand electron volts) determines the depth.
5. During predeposition, the substrate temperature (and, hence, the solid solubility) determines the concentration of dopant at the surface of the wafer.
6. The predeposition profile is determined by a combination of the time and temperature of the predeposition.
7. An oxide thickness of 0.12 microns will effectively mask a wafer against a boron diffusion for 1 hour and 1,100°C.
8. Seven methods of introducing dopant impurities into a silicon wafer are: solid source, liquid source, gaseous source, source wafers, chemical vapor deposition of oxide, spinning on doped oxide, and ion implantation.
9. The three variables which determine the junction depth during drive-in are the predeposition impurity concentration, the time, and the temperature.
10. The two most frequently used measurements are sheet resistivity and junction depth.
11. The constantly changing concentration profile leads to only an average resistivity measurement.
12. From Figure 5–10,
    a. The erfc of 3.5 is 0.7×10^{-6}.
    b. The number whose erfc is 3.5×10^{-3} is 2.1.
13. From Figure 5–12,
    a. The junction will be present at $\sim .3\mu$.
    b. The junction will be present at $\sim .13\mu$.

A4.6 Epitaxial Deposition

14. From Figure 5–13,
    a. The junction will be present at 1.75μ.
    b. The junction will be present at .45μ.

15. Graph data from problems 13 and 14.

16. Inversely since the resistivity decreases with increasing (Q).

17. As time progresses during a predeposition, the (Q) will increase and the resistivity will decrease.

18. Using Figure 5–17, the projected range and standard deviation are:
    a. 0.245 μm and 0.074 μm for 80 keV boron.
    b. 0.155 μm and 0.059 μm for 100 keV phosphorus.

19. Channeling is the creation of a bi-modal Gaussian distribution of an impurity during ion implantation. Channeling can be minimized by off-axis wafer orientation, implanting through an amorphous layer, such as silicon dioxide (SiO_2), or a combination of these two techniques.

20. For arsenic implanted into silicon at 200 keV,
    a. Using Figure 5–17, the projected range is 0.108 μm and the standard deviation is 0.043 μm.
    b. Using Figure 5–18, the mask thickness of silicon nitride to mask this implant is 0.185 μm.

A4.6 | Epitaxial Deposition

1. No. As long as the crystal structure of the substrate is continued through the deposited layer, it is an epitaxial layer.

2. A misalignment of 35 degrees will produce a maximum deposition rate.

3. From Figure 6–9, we see that above 4% HCl, a pitted surface will result.

4. a. From Figure 6–12, the maximum growth rate occurs when the mole fraction of $SiCl_4$ is 0.1.
   b. Silicon deposited using these growth conditions has poor crystal structure.

5. Nucleation sites are created on ⟨111⟩ silicon by first slicing the wafers 3 to 7 degrees off-axis and then etching the wafers to expose the sites. For ⟨100⟩ silicon, only the etch prior to deposition is needed.

6. Hydrogen reduction of silicon tetrachloride

$$SiCl_4 + 2H_2 \rightarrow Si + 4HCl$$

and pyrolysis of silane

$$SiH_4 \rightarrow Si + 2H_2$$

7. Using silane at 1050°C for 5 minutes yields an epitaxial layer of 10 micrometers from Figure 6–13.
8. a. Induction heating or RF—the RF energy is coupled directly into a carbon susceptor, which heats the wafers that are lying on it.
   b. UV—ultraviolet radiation from special bulbs heats the susceptor and the wafers by being directly absorbed.
9. The reacting species react less rapidly on a cold wall than on the hot susceptor, which means that there is less buildup on the wall.
10. a. Thickness
    b. Impurity concentration
    c. Crystal quality
11. $$d = \frac{n\lambda}{2} = \frac{(8)\ 0.3\ \mu m}{2} = 1.2\ \mu m$$
12. Epitaxial layer thickness can be determined by groove and stain or etch pit depth (see Section 6–3).
13. For etch pits 1.838 μm on a side, the thickness of the epi layer is 1.5 μm.

A4.7 Nonepitaxial Chemical Vapor Deposition

1. Hot-wall reactor—the reaction proceeds on the chamber wall as fast or faster than on the substrates. It is heated using thermal resistance heating.
2. a. Polycrystalline silicon
   b. Silicon dioxide
   c. Silicon nitride
3. Using Figure 7–6, the phosphorus concentration is 7×10^{20} atoms/cm^3.
4. The reaction chamber provides a controlled envelope around the reaction zone.
5. Reaction chamber
   Gas control section
   Time and sequence section
   Heat source for substrates
   Effluent handling system
6. Epitaxial growth is the special case of chemical vapor deposition during

which the grown layer assumes the same crystal orientation as the substrate.

7. $3SiH_4 + 4NH_3 \longrightarrow Si_3N_4 + 12H_2$

A4.8 | Metallization

1. See text.
2. See text.
3. E-beam
4. Planetary
5. Aluminum meets most of the metallization requirements.
6. To prevent reaction and electromigration, respectively.
7. The chamber, vacuum pumps, and monitor instrumentation.
8. Four deposition methods are filament, E-beam, flash, and induction evaporation.
9. A typical vacuum deposition cycle includes:

    Wafer clean and dry

    Wafer load

    Rough vacuum

    High vacuum

    Evaporate source

    Deposit

    Stop source

    Backfill

    Unload wafers

A4.9 | Device Processing: From Alloy to Sale

1. From Figure 9–5, the two compositions are:
   a. 22% Au, 78% Si
   b. 44% Au, 56% Si
2. From Figure 9–1, the composition is 11.3 atomic percent aluminum and 88.7 atomic percent silicon.

3. From Figures 9–1 and 9–5, we see that the aluminum-silicon eutectic temperature is higher.

4. a. Diamond scribing
   b. Laser scribing
   c. Sawing

5. a. TC bonding
   b. US bonding

6. The post-alloy probe step provides the designer with an indication of process variations.

7. Backside lapping and backside metal deposition prepare the backside of the wafer for die attach and also remove builtup contamination.

8. Gold has excellent soldering and thermal properties.

9. Nonfunctional die are inked during wafer sort.

10. Typical steps include scratch protection, backside preparation, wafer sort, device separation, die-attach, wire bonding, packaging, final test, and mark and pack.

11. Using Figure 9–2, for a p-type doping concentration of 2×10^{18} atoms/cm^3,
    a. R_c is 1.8×10^{-6} ohm-cm^2, and
    b. the total resistance of the two 4-micrometer $\times$ 6-micrometer contact windows is

    $$R_w = \frac{2(1.8 \times 10^{-6} \text{ ohm-cm}^2)}{(4 \times 10^{-4} \text{ cm})(6 \times 10^{-4} \text{ cm})} = 15.0 \text{ ohms}$$

    which is added to the 100-ohm resistor.

12. For a p-type base doping concentration of 6×10^{18} atoms/cm^3, the added base resistance for an R_c of 8×10^{-7} ohm-cm^2 (Figure 9–2) is

    $$R_{BC} = \frac{8 \times 10^{-7} \text{ ohm-cm}^2}{(4 \times 10^{-4} \text{ cm})(7 \times 10^{-4} \text{ cm})} = 2.86 \text{ ohms}$$

A4.10 Device and IC Technologies

1. Bipolar and MOS technologies

2. Bipolar technology has a minimum of seven masking steps while MOS technology has a minimum of five.

3. Shorting any two leads of a transistor together forms a diode between the two shorted leads and the third lead.

4. A pinched-base resistor usually has a much higher resistance per unit area.

5. See Figure 10–21.

6. Reduced saturation resistance

7. The gate oxide isolates the gate material from the channel.

8. The dielectric capacitor offers larger capacitance values per unit area than a junction capacitor. The dielectric capacitor also prevents conduction in both directions, while a junction capacitor conducts in one direction.

9. *N*-channel and *p*-channel transistors, resistors, diodes, and capacitors can be fabricated using MOS technology.

10. *NPN* transistors, *PNP* transistors, diodes, resistors, and capacitors can be fabricated using bipolar technology.

A4.11 The Wafer Fabrication Environment

1. Sodium

2. a. Deionization
   b. Reverse osmosis

3. a. Solvent clean
   b. The solvents remove organic contaminants such as waxes that may not react with the acids.

4. Ideally, the ceiling of a wafer fabrication area would consist entirely of laminar flow hoods. Unfortunately, their expense prohibits their use overall, except for the most sensitive processing stations.

5. Heat in H_2SO_4 to remove organics.
   Heat in dilute HF to clean oxides.
   Dip in dilute HF to clean oxides.
   Rinse in H_2O to remove acids.
   Dry to prepare for next step.

6. See Figure 11–2.

7. The major differences are the PH adjustment, the filtering, and the use of reverse osmosis.

8. Inert plastic piping

9. Copper tubing for oxygen and nitrogen, and stainless steel for others.
10. See text.
11. Graph Equation 11–1 on Figure 11–1. The 9-step yield for an area of 0.1 inch × 0.1 inch is

$$Y = \frac{1}{[1 + (1 \times 10^{-2})(5)]^9} = 0.645$$

or 64.5 percent. The yield decreases from 82 percent, so the net change (decrease) is 17.5 percent.

A4.12 Semiconductor Measurements

1. Using Equation 12–19 and a flat-band voltage shift $\Delta V_{FB} = 0.4$ volts from Figure 12–16, the mobile ion concentration is

$$N_I = \frac{1.2 \times 10^6 (277 \text{ pF})(0.4 \text{ V})}{(0.040 \text{ in})^2} = 8.31 \times 10^{10} \text{ ions/cm}^2$$

2. Using Equation 12–18, the oxide thickness is just

$$t_f = \frac{1.72 \times 10^8 (0.040 \text{ in})^2}{(277 \text{ pF})} = 994 \text{ angstroms}$$

3. For $t = s$ in Figure 12–2, the correction factor, A, is just 0.92. From Equation 12–2,

$$R_s = 4.53(45)(0.92) = 187.5 \text{ ohms per square.}$$

4. Using Equation 12–3, for a 12-micrometer (12×10^{-4} cm) uniformly doped layer, the resistivity is

$$\rho = (187.5 \text{ }\Omega/\square)(12 \times 10^{-4} \text{ cm}) = 0.26 \text{ ohm-cm.}$$

5. Four commonly used diffusion profile measurements are capacitance-voltage (CV), spreading resistance, secondary ion mass spectroscope (SIMS), and differential conductivity.

6. From Figure 12–16, the flat-band voltage, ΔV_{FB}, is 0.5 volts, and the flat-band capacitance is

$$0.74 \times 277 \text{ pF} = 205 \text{ pF.}$$

A4.13 Advanced Silicon Technology

1. X-rays or electron beams
2. Larger wafers produce more devices for the same fabrication sequence. Practical considerations such as furnace size limit wafer size. Wafers as large as 10 to 12 inches have been manufactured.
3. The minimum dimensions that can be transferred is 2 to 4 times the wavelength of the light or 1.2 to 2.4 μm.
4. 1.0 to 2.0 μm
5. Electron beams may be used to obtain magnification, and a beam of electrons may be deflected since electrons are charged particles.
6. See Figure 13–3 and 13–4.
7. CCDs may be used to manufacture imaging devices such as television cameras.
8. Temperature, pressure, and magnetic fields

A4.14 Nonsilicon Technology

1. An LED display requires more power because it actually emits light as opposed to just reflecting or transmitting it.
2. Quartz is a piezoelectric material which means that it vibrates when a voltage is applied.
3. Thin-film hybrid circuits use thin layers of vacuum-deposited material while thick-film hybrid circuits are formed by screening a layer of paste on a substrate and firing it at an elevated temperature.
4.
   a. Epitaxial deposition of the material that will serve as the light emitting material
   b. Oxide deposition using CVD
   c. Image transfer to the front surface using photolithography
   d. Form the diode using a high temperature diffusion
   e. Form ohmic contacts to the diode
   f. Separate the substrate into discrete diodes
   g. Test and package the devices
5. Laser diodes produce light with only one wavelength, while LEDs produce light across a range.
6. The mobility of the carriers is 2 to 3 times higher in GaAs than in silicon.

7. Gallium arsenide does not form an oxide, and it decomposes at high temperatures. For these reasons, conventional IC fabrication techniques do not work.

8. Bubble memories are attractive in electrically noisy environments or where the power may be interrupted.

APPENDIX 5
Glossary

Acceptors: Atoms that contribute holes to the electrical conductivity of a semiconductor. For silicon, these elements reside in column III of the periodic table.

Aligner: The piece of equipment that is used to position the mask and the wafer for exposure in the photomasking process.

Alloy: In semiconductor processing, the alloy step causes the interdiffusion of the semiconductor and the material on top of it, forming an ohmic contact between them.

Anisotropic etch: An etch that removes material more rapidly in one direction.

Aluminum: The metal most often used in semiconductor processing to form the interconnects between the devices on an integrated circuit chip. It is frequently deposited by evaporation and sputtering.

Angle lap: A method for magnifying the depth of a junction by cutting (lapping) through it at an angle away from the perpendicular.

Anneal: A high-temperature processing step which is used to minimize surface effects in devices by relieving stress or after ion implantation to repair the damage to the crystal lattice.

Anodic oxidation: The growth of a layer of silicon dioxide at low temperature by using the silicon wafer as an anode and placing a voltage between the wafer and a cathode in an electrolyte.

Antimony trioxide (Sb_2O_3): A solid that is often used as a source of antimony for doping silicon.

Argon: An inert gas frequently used as the ambient gas during the silicon crystal growing process.

Arsenic: The n-type dopant used for the subcollector and emitter diffusions in bipolar npn transistors, and the source and drain of n-channel MOS integrated circuits.

Arsenic trioxide (As_2O_3): A solid that is used as a source of arsenic for doping silicon.

Arsine (AsH_3): A gas that is often used as a source of arsenic for doping silicon.

Base: In a bipolar transistor, the terminal that controls the current flow.

Bipolar: A type of transistor where the flow of both conduction electrons and holes determines the device characteristics.

Boat: 1. Pieces of quartz joined together to form a supporting structure during high-temperature processing. 2. A Teflon or plastic assemblage used to hold wafers during wet processing.

Boron: A p-type dopant commonly used for the isolation and base diffusion in standard bipolar technology and the source and drain of p-channel MOS integrated circuits.

Boron nitride (BN): A solid wafer source that is used as a source of boron for silicon doping.

Boron tribromide (BBr_3): A liquid that is used as a source of boron for doping silicon.

Boron trichloride (BCl_3): A gas that is often used as a source of boron for silicon doping.

Bubbler: A closed vessel containing a liquid such as water or a dopant source. A carrier gas bubbles through the liquid carrying its vapor into the furnace.

Burnt hydrogen: See pyrogenic steam.

Capacitance-voltage (CV): A measurement technique used to determine doping concentrations, contamination levels, doping profiles, and oxide thickness.

Capacitor: A circuit element formed by placing an insulating layer between two conducting layers. This structure is often fabricated in the manufacture of integrated circuits.

Carrier gas: A non-reactive gas that flows through a furnace tube or a reaction chamber to maintain uniform conditions for the reactive species.

Channeling: The process whereby some portion of the ion beam penetrates to a much greater distance than the main portion during ion implantation. Channeling can be minimized by a combination of off-axis bombardment and implanting through an amorphous layer of SiO_2.

Charge-coupled device (CCD): A device that works by moving discrete packets of charge under the influence of electric fields provided by surface electrodes.

Chemical vapor deposition (CVD): The process of forming a stable compound on a heated substrate by thermally reacting or decomposing gaseous compounds on the surface.

Collector: In a bipolar transistor, the terminal to which the carriers flow.

Complementary MOS (CMOS): MOS technology that uses both *n*-channel and *p*-channel MOS transistors. This technology has low power dissipation.

Conduction electron: An electron that has broken free from an atom in the crystal structure of a semiconductor, leaving it free to move in an applied electric field. It has a negative charge.

Conductivity (electrical): The inverse of resistivity.

Conductivity (thermal): The ease with which heat flows through a material.

Contamination: A general term used to describe unwanted material that adversely affects the physical or electrical characteristics of a semiconductor wafer and the resulting die yield from that wafer.

Covalent bonding: The bonding between atoms that results from the sharing of electrons.

Dehydration bake: A bake performed at a temperature above 100°C to evaporate any moisture from the surface of a wafer.

Diborane (B_2H_6): A gas that is used as a source of boron for doping silicon.

Dichlorosilane (SiH_2Cl_2): A gas that is used as a source of silicon in CVD processes.

Diffusion: 1. The movement of particles away from regions of high concentration caused by the random thermal motion of atoms and molecules. 2. A process used in the production of semiconductors which introduces minute amounts of impurities into a substrate material such as silicon or germanium. The process, which is very dependent on time and temperature, permits the impurity to spread into the substrate.

Diode: A two-terminal electronic component consisting of a *pn*-junction. This junction allows current to flow in one direction but not in the other.

Dislocations: Imperfections in the local crystal structure of a silicon wafer.

Donors: Atoms that contribute conduction electrons to the electrical conductivity of a semiconductor. For silicon, these elements reside in column V of the periodic table.

Drain: In an MOS transistor, the terminal to which the carriers flow.

Drift: The movement of mobile carriers (holes or conduction electrons) in an electric field.

Drive-in: The redistribution of the dopant introduced during the predeposition step to obtain the final position of the dopant in the wafer.

E-beam: See electron beam.

Effluent: Waste gases or liquids that are a byproduct of the semiconductor manufacturing process. They are cleaned of any harmful or reactive components and then disposed of properly.

Electromigration: The movement (migration) of atoms in the metallization caused by the flow of electrical current.

Electron beam: A type of evaporation that uses the energy of a focused electron beam to provide the required heat.

Ellipsometer: An instrument utilizing polarized laser light to measure transparent film thickness.

Emitter: In a bipolar transistor, the terminal from which the carriers flow.

Epitaxial deposition (epitaxy): The deposition of a single crystal layer on a substrate so the crystal structure of the deposited layer matches the crystal structure of the substrate.

Eutectic temperature: This is the lowest temperature at which a molten solution of a mixture of two materials can exist. The eutectic temperature for an aluminum-silicon mixture is 577°C.

Evaporation: A process step that uses heat to evaporate a material from a source and deposit it on wafers. Both electron-beam and filament evaporation are common in semiconductor processing.

Filament: A coiled piece of wire that is loaded with a material to be evaporated and heated by passing current through it.

Flat: That portion of the silicon ingot which is ground along the length of silicon ingot. It is later used as a reference once the ingot is sliced into wafers.

Flatness: A reference to the relative deviations in the overall topography of the surface of a silicon wafer. It can be measured with the wafer either unrestrained or under a vacuum applied with a chuck to the back side of the wafer. Typical flatness deviations are usually less than 3 to 5 micrometers.

Four-point probe: A piece of electronic test equipment used to determine the sheet resistivity of a predeposition or a diffusion.

Furnace: A piece of equipment containing a resistance-heated element and a temperature controller. It is used to maintain a region of constant temperature with a controlled atmosphere for the processing of semiconductor devices.

Gate: In an MOS transistor, the terminal that controls the current flow.

Groove and stain: A technique used to determine the thickness of a doped region of silicon on an oppositely doped substrate (n on p, or p on n). A groove is "lapped" through the top layer and a stain is used to delineate the junction. An optical technique gives the layer thickness.

Hall effect: The voltage that is produced perpendicular to the direction of current flow when a magnetic field is applied to a conducting region of a semiconductor.

Hard bake: The higher temperature bake step after the expose step that increases the adhesion of the resist at edges.

Hole: The electrically-charged mobile carrier formed by the absence of an electron in the atomic structure of a semiconductor. It has a positive charge.

Hydrofluoric acid (HF): A strong acid used to etch silicon dioxide. It is usually used diluted and/or buffered.

Inert gases: The elements that are in the column on the right side of the periodic table. These elements have a complete set of electrons in their outer shell and, therefore, do not react chemically.

Impurity (also see "dopant"): A foreign material or element that finds its way into the silicon wafer. The impurity may be necessary as in the case of dopants, or it may be present due to improper wafer handling or processing.

Ingot: The long, single-crystal rod of silicon which is the result of the crystal-growing process.

Interferometer: An instrument used to determine the thickness of a transparent layer by observing the number of interference fringes caused by the interaction of the light waves reflected from the upper and lower surfaces of the film.

Intrinsic: Semiconductor material containing no impurities of any type.

Ion: An atom that has either gained or lost electrons, making it a charged particle (positive if it has lost electrons and negative if it has gained electrons).

Ion implantation: The process of introducing dopant atoms into the surface of a silicon wafer by accelerating the ionized dopant atoms in a large electric field. The wafer is then bombarded by these high energy dopant ions causing them to penetrate into the exposed portions of the wafer.

Ionic bonding: The bonding between atoms that occurs when one atom borrows an electron from another, making each atom a charged "ion." The electrostatic attraction between the two charged atoms bonds them together.

Isotopes: Atoms that are the same element but have different numbers of neutrons in their nucleus, giving them different atomic weights.

Isotropic etch: An etch that removes material uniformly in all directions.

Junction (junction depth): the place at which the conductivity type of a material changes from p-type to n-type or vice versa.

Lap and stain: See "groove and stain."

Lapping: A free-abrasive mechanical machining process which is used to planarize both sides of a silicon wafer as well as to finish the wafer prior to polishing.

Laser diode: A light-emitting diode whose light output is at a single frequency.

Light-emitting diode (LED): A semiconductor diode (usually a non-silicon diode) that emits light when current flows through it in the forward direction.

Liquid crystal display (LCD): A display that uses an organic crystal (liquid crystal) whose light reflecting properties change when a voltage is applied. Electrodes on the front and back surface of the display panel are used to apply a voltage.

Metal oxide semiconductor (MOS): A type of transistor where a voltage on a conductive "gate" region produces a conductive channel between two doped regions, allowing current to flow.

Metallization: The layer of high-conductivity material (a metal or alloy) used to interconnect devices on a chip. Aluminum is the most frequently used with silicon.

Microspectrophotometer: An instrument utilizing multiple wavelength light to determine the thickness of thin transparent films.

Miller indices: A method used to describe the orientation of the crystal planes within a material such as crystalline silicon. Frequently used silicon crystal planes are the $\langle 111 \rangle$ and $\langle 100 \rangle$ planes.

Mobility: The ease with which carriers (holes or conduction electrons) move when an electric field is applied.

N-channel MOS: MOS transistors with n-type source and drain regions. The "channel" allowing current to flow is also n-type or n-channel.

N-type: A shortened version of "negative-type," indicating that the semiconductor contains more conduction electrons than holes.

Nitrogen (N_2): A gas that seldom reacts with other materials. It is often used as a carrier gas for chemicals in semiconductor processing.

Ohmic: A term used to denote a linear relationship between the voltage across a region and the current through it. An ohmic contact has this linear relationship, but hopefully the resistance is low.

Oxygen (O_2): A gas used in semiconductor processing to oxidize silicon, to form vapor-deposited oxide, and for other processing steps.

P-channel MOS: MOS transistors with p-type source and drain regions. The "channel" allowing current to flow is also p-type or p-channel.

P-type: A shortened version of "positive-type," indicating that the semiconductor contains more holes than conduction electrons.

Passivation: A layer of material covering a device or circuit to prevent unwanted contaminants from the environment from altering its performance.

Periodic table of the elements: The arrangement of elements into a row and column format that groups elements with similar atomic structure. This arrangement provides information about the chemical behavior of the elements.

Phosphine (PH_3): A gas that is often used as a source of phosphorus for doping silicon.

Phosphorus: The *n*-type dopant commonly used for the emitter diffusion in standard bipolar integrated circuit technology and the *n*-channel source and drain of MOS integrated circuits.

Phosphorus oxychloride ($POCl_3$): A liquid that is used as a source of phosphorus for doping silicon.

Phosphorus pentoxide (P_2O_5): A solid that is used as a source of phosphorus for doping silicon.

Photolithography: In semiconductor manufacturing, this process step transfers an image from a mask to the surface of a wafer through the use of light.

Photoresist: A photo- or light-sensitive, etch-resistant material used for transferring an image to the surface of a wafer.

Polycrystalline silicon (polysilicon): Silicon composed of many (poly) crystals. Raw silicon comes in ingots of poly prior to crystal growth. Poly may be deposited accidentally during epitaxial deposition by depositing it too fast or at too low a temperature. CVD of poly silicon usually occurs on a layer of silicon dioxide.

Predeposition (also called predep): The process step during which a carefully controlled amount of dopant is introduced into the crystal structure of a semiconductor using solid state diffusion techniques.

Preform: A small piece of special composition material that will adhere to both the die and the package when heated.

Priming: Coating the surface of a wafer with a chemical that increases the adhesion of photoresist on the surface.

Prism coupler: An instrument utilizing the tunneling of photons from the base of a specially constructed prism in contact with a thin transparent film to measure its thickness.

Profilometer: An instrument utilizing the movement of a diamond stylus over a substrate to measure heights of various features on a processed silicon wafer.

Pyrogenic steam: Steam (water vapor) formed by the reaction of hydrogen and oxygen. Steam formed in this fashion is usually used to oxidize wafers.

Quartz: The term used to denote the silicon dioxide used in the manufacture of diffusion tubes, boats, etc. for use in high temperature processing steps.

Radio frequency (RF): The energy medium used to heat the susceptor in most epitaxial reactors and in crystal-growing furnaces. Radio frequency means that the energy is transferred at a frequency near the radio transmitting band.

Reactor: A piece of equipment used for the deposition of a layer of material used in semiconductor processing. Common types of reactors are epitaxial reactors, vapor reactors, and nitride reactors.

Reflectivity: A measure of the ability of a surface or substrate to return incident illumination. A silicon wafer, for example, has a reflectivity of 33 to 35 percent for visible light.

Resistivity (ohm-cm): A measure of the difficulty in moving electrons (and holes) through a material such as silicon. When dopants are added to the material, it is an indication of the dopant concentration.

Resistor: A circuit element that allows current to flow when a voltage is applied.

Resistance (R): The opposition to the motion of charged carriers (holes or conduction electrons) that characterize a volume of a material. Its units are "ohms." The linear relationship between the current and the voltage is the resistance ($V = RI$).

Scribe: The process of forming a shallow trench either mechanically or with a laser between each die on a finished wafer. The scribe lines are then used to initiate breaking the wafer into individual die.

Sheet resistance (R_s): A measurement with dimensions of ohms per square that is frequently used in evaluating predepositions and drive-ins. It is related to the number of n-type donor or p-type acceptor donor or acceptor atoms in a semiconductor.

Silane (SiH_4): A gas used as a source in the CVD of silicon and silicon compounds.

Silicon (Si): The group IV element used for fabricating diodes, transistors, and integrated circuits.

Silicon dioxide: The oxide of silicon that is used—either deposited or thermally grown—as an insulating layer and as a barrier to unwanted impurities.

Silicon pyrophosphate (SiP_2O_7): A solid wafer source used for n-type phosphorus doping of silicon.

Silicon tetrachloride ($SiCl_4$): A gas that reacts with hydrogen producing silicon and hydrogen chloride gas. It is often used to deposit epitaxial silicon.

Sintering: See alloy.

Slip: A type of silicon crystal defect which results in one part of the crystal plane being sheared with respect to the neighboring portion of the crystal.

Appendix 5: Glossary

Sodium: A highly mobile metal atom that drifts through silicon dioxide during semiconductor processing, adversely affecting device performance and yield.

Soft bake: The lower temperature bake step, prior to the expose step, that evaporates the solvent from the photoresist layer.

Source: In an MOS transistor, the terminal from which the carriers flow.

Spinner: The piece of equipment that is used to coat a wafer with photoresist or sometimes to develop the photoresist.

Spreading resistance: A measurement used in conjunction with angle-lapping to determine the doping profile in a diffused region.

Sputtering: A method of depositing a film of material on a desired object. A target of the desired material is bombarded with RF-excited ions which knock atoms from the target and deposit them on the object to be coated.

Susceptor: A flat slab of material (usually graphite) on which wafers are heated by induction during high-temperature deposition processes such as epitaxial growth or nitride deposition.

Transistor: A three-terminal circuit element manufactured using semiconductor material. The transistor provides signal amplification.

Trichlorosilane ($SiHCl_3$): A liquid at room temperature containing silicon. The liquid is easily purified to semiconductor standards by fractionation (distillation) procedures. When trichlorosilane is heated in the presence of hydrogen, polysilicon is produced with hydrochloric acid as a byproduct.

Tube (also see "furnace"): A cylindrical piece of quartz with fittings on one or both ends. It is placed in a furnace to provide a contamination-free, controlled atmosphere during several of the processing steps.

Vacuum wand: A hand-held apparatus for picking up wafers by their back surface using a vacuum.

Viscosity: A measure of the ease with which a liquid flows. (Higher viscosity liquids flow more slowly.)

Wafer: A round thin slice of a semiconductor material. Often used when referring to a wafer of silicon.

Index

acceptors, 8
aligner, 58, 59
alloy, 149
aluminum, 136
angle lap, 117
anisotropic etch, 61, 62
anneal, 149
anodic oxidation, 48, 49
antimony, 8
antimony trioxide, 80
argon, 23
arsenic, 8, 112
arsenic trioxide, 80
arsine, 80, 112

base, 162
bipolar, 159, 161, 172
boat, 35
boron, 8, 47, 81
boron nitride (BN), 81
boron tribromide (BBr_3), 81
boron trichloride (BCl_3), 81
bubbler, 36

capacitance-voltage (CV), 41, 119, 189
capacitor, 167, 169
carrier gas, 109
channeling, 97
charged-couple device (CCD), 211, 212, 213
chemical vapor deposition (CVD), 123
chemical vapor deposition reactor, 21
collector, 162

complementary error function, 83, 85
complementary MOS (CMOS), 105, 170
conduction electron, 6, 7
conductivity, 8, 12
contact resistance, 151
contamination, 175
copper, 136
correction factor, small sample geometries, 187
correction factor, thick materials, 185, 186
coulomb, 12
covalent bonding, 4
crucible, crystal growth, 22
crystal defects, 29
crystal growth, Czochralski, 22
crystal growth, float zone, 22
crystal orientation, 27
current gain, 161

defect density, 175
dehydration bake, 55
depletion mode, 217
diamond scribing, 155
diborane (B_2H_6), 112, 117
dichlorosilane (SiH_2Cl_2), 111
differential conductivity measurement, 192, 193
diffraction, 208
diffraction, X-ray, 26
diffusion, 14, 34, 42, 71
diffusion coefficient, 72, 82
diode, 103, 162
dislocations, 29

distribution coefficient, 25
donors, 8
dopant, 7, 73
dopant redistribution, 46
drain, 168
drift, 14
drive-in, 73

E-beam, 142, 144
effluent, 112, 113
electromigration, 136
electron beam, 142, 144
electron beam exposure, 209
electrons, 1
ellipsometer, 40, 196, 197
emitter, 162
epitaxial deposition (epitaxy), 103, 106, 123, 211
etching, 52, 61, 62, 211
eutectic die attach, 156
eutectic temperature, 150
evaporation, 142

filament, 143
filament evaporation, 142, 143
flash evaporation, 142, 144
flat, 30
flatness, 27
four-point probe, 10, 185, 186
furnace, 35

gallium arsenide, 217
gate, 169
gate oxide, 40
Gaussian distribution, 86, 87
germanium, 5

263

groove and stain, 117

Hall effect, 213
hard bake, 61
HCl, 41, 110, 113
HEPA filter, 182
hole, 6, 7
hot probe, 16
hybrid technology, 219
hydrofluoric acid (HF), 112

impurity, 7
induction evaporation, 142
inert gases, 3
infrared interferometer, 193
infrared spectrophotometer, 193
ingot, 25
insulator, 4
interferometer, 198
intrinsic, 7
ion, 1
ion exchange, 179, 180
ion implantation, 94
ionic bonding, 4
isotopes, 2
isotropic etch, 61, 62

Josephson junction, 218
junction, 84, 188

laminar flow hood, 181
lap and stain, 117, 189
lapping, 26
laser diode, 215
laser inspection, 182, 183
laser scribing, 155
light-emitting diode (LED), 103, 215, 216
liquid crystal display (LCD), 216

magnetic bubble (domain) devices, 218
magnetic field sensor, 213
magnetron sputtering, 145
Mendeleev, Dimitri Ivanovich, 4
mercury arc lamp, 51, 53, 60
mercury probe, 194, 195
metal, 4
metal oxide semiconductor (MOS), 159, 168, 170, 172
metallization, 135
microspectrophotometer, 199
Miller indices, 27
mobility, 13

N-channel MOS, 168, 170

N-type, 7
negative resist, 53, 60
neutrons, 1
nitrogen, 35, 36
nucleus, 1

optical dosimeter, 203, 204
optical interferometer, 198
oxidation, 33, 34, 35, 36, 43, 55, 211
oxidation, anodic, 48, 49
oxidation, high pressure, 47
oxidizing species, 34
oxygen, 34, 36, 66

P-channel MOS, 170
P-type, 7
particle detector, 213
passivation, 130
periodic table of the elements, 2, 3
phosphine (PH_3), 112, 116
phosphorus, 8, 46, 80
phosphorus oxychloride ($POCl_3$), 80
phosphorus pentoxide (P_2O_5), 80
photolithography, 51
photomask, 66
photoresist, 51, 52, 53, 55
polycrystalline silicon, 19, 128
positive resist, 53, 60
predeposition (predep), 72
preform, 156
preform die attach, 156
pressure transducer, 213
priming, 55
prism coupler, 196, 200
profilometer, 205
protons, 1
pyrogenic steam, 37

quartz, 35, 106, 218

radio frequency (RF), 107, 126
reactor, 105, 106, 107
reflectivity, 193
resistance, 8, 9
resistivity, 8, 12, 187
resistivity, average, 188, 189
resistor, 164
reverse osmosis (RO), 179, 180

sawing, 155
Schottky diode, 194
scribe, 155

seed crystal, 23
semiconductor, 5
sheet resistance, 10, 11, 185, 186
silane (SiH_4), 111, 115
silicon, 5
silicon, polycrystalline, 19, 128
silicon dioxide, 19, 33, 40, 46, 128
silicon nitride, 128
silicon preparation, 19
silicon pyrophosphate (SiP_2O_7), 80
silicon tetrachloride ($SiCl_4$), 111, 113, 114, 115
SIMS, 192, 193
slip, 30
sodium, 35, 41
soft bake, 57, 61
soft solder die attach, 156
solid solubility, 73, 74
source, 168
spiking, 136
spinner, 55
spreading resistance, 119, 189, 191
sputtering, 142, 144
superconductor, 218
surface profilometer, 40
susceptor, 107, 108

temperature sensors, 213
thermocompression bonding, 157
thermosonic bonding, 157
thick film, 219
thin film, 219
transistor, 103, 161
trichlorosilane ($SiHCl_3$), 20, 111
tube, 78

ultrasonic bonding, 157
ultraviolet heating, 126
ultraviolet light, 51, 109

vacuum gauges, 141
vacuum pumps, 138
vacuum wand, 113
vapor pressure, 37
viscosity, 54

wafer, 19
water vapor, 34, 36, 37

X-ray exposure, 209

yield, 175

Zener diode, 162